Synthese Library

Studies in Epistemology, Logic, Methodology, and Philosophy of Science

Volume 495

The aim of *Synthese Library* is to provide a forum for the best current work in the methodology and philosophy of science and in epistemology, all broadly understood. A wide variety of different approaches have traditionally been represented in the Library, and every effort is made to maintain this variety, not for its own sake, but because we believe that there are many fruitful and illuminating approaches to the philosophy of science and related disciplines.

Special attention is paid to methodological studies which illustrate the interplay of empirical and philosophical viewpoints and to contributions to the formal (logical, set-theoretical, mathematical, information-theoretical, decision-theoretical, etc.) methodology of empirical sciences. Likewise, the applications of logical methods to epistemology as well as philosophically and methodologically relevant studies in logic are strongly encouraged. The emphasis on logic will be tempered by interest in the psychological, historical, and sociological aspects of science. In addition to monographs *Synthese Library* publishcs thematically unified anthologies and edited volumes with a well-defined topical focus inside the aim and scope of the book series. The contributions in the volumes are expected to be focused and structurally organized in accordance with the central theme(s), and should be tied together by an extensive editorial introduction or set of introductions if the volume is divided into parts. An extensive bibliography and index are mandatory.

Gal Yehezkel

Conceptual Principles of Natural Philosophy

An Investigation into Time, Space, and the Laws of Nature

Springer

Gal Yehezkel
Department of Liberal Arts and Sciences
The Sapir Academic College
D.N. Hof Ashkelon, Israel

ISSN 0166-6991 ISSN 2542-8292 (electronic)
Synthese Library
ISBN 978-3-031-82107-3 ISBN 978-3-031-82108-0 (eBook)
https://doi.org/10.1007/978-3-031-82108-0

This Springer imprint is published by the registered company Springer Nature Switzerland AG
The registered company address is: Gewerbestrasse 11, 6330 Cham, Switzerland

To Neomi, Ofek, Arava, and Shahak

Acknowledgments

This book encapsulates years of inquiry. My sincere thanks go to all those who read and commented on earlier drafts, engaged in conversations on its ideas at different stages, taught me philosophy, and helped in various ways along the journey. Special thanks to Robert Albin, Dalia Drai, Jerome (Yehuda) Gellman, Yuval Lurie, and Haim Marantz.

Most of all, I am grateful to my family: my wife, Neomi, for her patience, understanding, and support, and our children, Ofek, Arava, and Shahak, who always reminded me whom I am writing for. This book is dedicated to them.

Contents

Chapter 1
Introduction

Abstract My purpose in this book is to put forward a philosophical research project, which is complementary to the empirical study of nature. The purpose of this project is to describe the conceptual structure of nature, thus providing a general framework for understanding nature, which I term "internal relationalism." My inquiry is likened to an attempt to establish nature from nothing, gradually introducing the key features of existence, such as time, the laws of nature, and finally space, thus explaining both their essence and their affinity.

Keywords Conceptual analysis · Conceptual structure · Natural philosophy

When Isaac Newton published his *Philosophiæ Naturalis Principia Mathematica* in 1687, the exploration of nature was considered a philosophical task, as the name of his book suggests. However, with time, and largely inspired by Newton's achievements, an independent empirical science was established—physics—whose end is to describe the fundamental mathematical structure of the totality of all things that exist, that is, nature (see also: Gascoigne, 2006; Kerszberg, 2006).

While the empirical exploration of nature progressed and grew more and more successful, the same cannot be said for the a priori study of nature. One hundred years after the publication of the *Principia*, in his preface to the second edition of the *Critique of Pure Reason*, Immanuel Kant noted that metaphysics, as an a priori exploration of existence, of which natural philosophy is a branch, has yet to become a science, that is, to achieve a state of agreement and progress (Kant, 1929, p. Bxiv). Unfortunately, Kant's attempt to lay the metaphysical foundation of the exploration of nature has likewise failed to establish natural philosophy as a science (Kant, 2002). The same can be said today, more than 300 years after the publication of the *Principia*. Although current metaphysics is occupied with many of the subject matters of the traditional philosophy of nature, that is, time, space, and the laws of nature, it has failed to establish natural philosophy as a science, complementary to the empirical investigation of nature. Nowadays it seems clear to anyone interested in understanding nature that physics, as an empirical science, rather than philosophy, is the path to take. Indeed, for generations now, the aspiration of studying

G. Yehezkel, *Conceptual Principles of Natural Philosophy*, Synthese Library 495, https://doi.org/10.1007/978-3-031-82108-0_1

nature through an a priori method of inquiry, whose only guide and tools are conceptual, is considered suspect, if not downright preposterous.

This is unfortunate. For, notwithstanding empirical science's progress and accomplishments in formulating a mathematical description of phenomena, there seems to be a limit to what can be achieved by such an investigation. While mathematics can describe the relations that obtain between key features of nature, such as time and space, it is unclear whether it can offer a true insight into the essence of these entities. Can, for example, time dilation, which follows from Albert Einstein's Special Theory of Relativity, truly offer an insight into the nature of time, if it follows from an experimental fact that was elevated into a postulate, that is, the principle of the invariant speed of light, which therefore has no further explanation? Notice that even if such an explanation can be given, it seems that the empirical research of nature must ultimately rely on unexplainable facts, for it cannot explain how things must be, but only describe phenomena as it finds them. How can empirical science offer a true understanding of the fundamental features of nature—whether they are time, space, or other more basic elements—if it cannot explain *why* these, rather than others, are the features that define nature? Can there be, for example, an atemporal universe? Surely all that an empirical inquiry can do is explore how things are contingently—it can never discover how things necessarily are. Notice that this is true even if it is accepted that a posteriori knowledge can be interpreted as expressing necessary truths, as suggested, for example, by Kripke (1980). Only philosophical considerations can bestow this status on an empirical finding. Moreover, how can empirical research, which sets as its end the discovery of the laws of nature, explain why nature is subject to any contingent laws at all? In fact, as demonstrated by David Hume's problem of induction, empirical research cannot even justify the presumption that such laws obtain in nature, on which this research project ultimately relies (Hume, 1978, p. 89). Finally, an empirical investigation cannot answer the most fundamental question: "Why is there something rather than nothing?" For if the most it can offer is an equation that describes the mathematical relations that obtain between the fundamental features of nature, even if physicists succeed in formulating a final and comprehensive "theory of everything," in the words of Stephen Hawking, the question will nevertheless remain: "What is it that breathes fire into the equations and makes a universe for them to describe?" (Hawking, 1998, p. 178).

The fact that physics cannot answer these essential questions shows that the empirical exploration of nature, notwithstanding its achievements, cannot offer us a thorough understanding of nature. A metaphysical science of natural philosophy, on the other hand, should and must be able to answer all the questions I have listed above. However, I wish to emphasize that it is not my intention, by outlining some of the limitations of empirical inquiry, to discredit the empirical exploration of nature. As shall become evident in what follows, according to the conception of nature I am putting forward in this book, the empirical study of nature is indispensable. What I argue is that there is also a need for an a priori science of nature, as a complementary research project, which could also contribute to the progress of the empirical research of nature (see also Maxwell (2012)).

Natural philosophy, as an a priori science, is hence necessary for a thorough understanding of nature. Furthermore, while the beginning of the twentieth century saw great revolution and advancement in the foundations of physics, this progress has slowed down in the last decades, if not completely stopped (Smolin, 2006, p. xi). It is interesting to note that this revolution in physics was influenced by philosophical ideas. Einstein, for example, was inspired in developing his theory of relativity by philosophical considerations, whose roots are found in the debate between Newton and Leibniz on whether space and time are absolute or relational (Agassi, 1969). In this context, it is noteworthy that this debate has its roots as far back in history as the seventeenth century, when exploring nature was still considered a philosophical endeavor. In the second half of the twentieth century, on the other hand, it suffices to quote Steven Weinberg, who wrote "I know of no one who has participated actively in the advance of physics in the post-war period whose research has been significantly helped by the work of philosophers" (Weinberg, 1993, p. 134). Indeed, some have recognized the need to "reinvent, the vanished genre of natural philosophy" and sought to "breathe new life" into this discipline (Unger & Smolin, 2014, pp. xvii–xviii).

I believe that philosophy can contribute to the understanding of nature, as well as to the advancement of the empirical exploration of nature—not as a philosophy of science, which follow in the footsteps of the empirical sciences, but rather as an independent and complementary science of nature. I am well aware of the skepticism that the declaration of such an aspiration will arouse. However, I wish to convince the reader that the a priori exploration of nature is possible, and that the reason for past failures is rooted in the difficulty of this of type research, rather than in its impossibility. Nevertheless, I should stress that I do not intend to argue for its possibility. My aim in this book is to demonstrate that the a priori study of nature is a viable research project, by establishing its foundations, articulating its principles, and implementing these tools in an a priori investigation of nature, thereby formulating an answer to all the above-listed queries, as well as many other fundamental questions about nature.

Physics tries to describe the mathematical structure of nature—in the words of Galileo, "…this all-encompassing book […] that is the universe […] is written in mathematical language" (Galilei, 2008, p. 183). The philosophy of nature, on the other hand, attempts to describe the conceptual structure of nature. While empirical research starts with investigating isolated individual phenomena and tries to elevate these findings into general principles that could expand our understanding of nature, an a priori investigation begins by formulating the most general principles of existence. Based on these principles, it advances to construct the more specific features of nature.

The objective of this book is therefore to provide a general framework for understanding nature. Just as mathematical relations describe the relations between key features of nature, conceptual connections define the most basic structure of existence. The role of natural philosophy is to explore this structure and show how it constitutes the fundamental features of nature. In an earlier book, I attempted to uncover the conceptual structure that underlies our actual conceptual scheme

(Yehezkel, 2014). In this book, I attempt to reconstruct the conceptual structure that constitutes nature from nothing, by developing the conceptual structure that is needed to give meaning to the concept of existence. Thus, I gradually introduce the key features of nature, time, the laws of nature, and finally space, and explain their affinity. This can be seen as an attempt to constitute nature from nothing, thereby exposing its necessary structure and its most fundamental features.

I should stress that I am not arguing that the conceptual structures identified through linguistic analysis are sufficient to account for the full complexity of temporal and spatial *phenomena*. In fact, the conceptual and the empirical—or phenomenological—aspects should be clearly distinguished. The complex connection between the conceptual (and therefore necessary) and the empirical (and therefore contingent) will be clarified in the course of my analysis. Meanwhile, I remind the reader that just as conceptual considerations alone cannot explain the contingent, the empirical cannot explain the necessary. Experience can only teach us how things are, not how they must be. Hence, the integrity of conceptual analysis does not allow for any empirical considerations.

Furthermore, this implies that there can be no contradiction between the empirical and the conceptual. Integrating the conceptual with the empirical is the goal of human endeavor in understanding nature. This integration will be achieved when the conclusions of empirical research, which seeks to identify the general principles underlying phenomena, overlap with the conclusions of conceptual research, which begins with the most general principles of nature and explores their implications. This claim will be further discussed and explained in the conclusion of this book.

The general outline of this book is as follows. I begin (in Chap. 2) by developing the conceptual tools and principles that are needed for such an inquiry. I implement these tools (in Chap. 3) in an analysis of the concept of existence. This analysis concludes that only the things that are described by contingent propositions can be meaningfully said to exist. In light of this conclusion, I show (in Chap. 4) that the temporal distinctions between past, present, and future are necessary features of nature, and that they imply that existence is determined in relation to something internal to nature—a point of view from which nature is described. Next (in Chap. 5) I reveal the role and structure of the laws of nature. The conclusions of this inquiry are used (in Chap. 6) for an objective reconstructing of the temporal and spatial dimensions and explaining their affinity. This constitutes the most fundamental conceptual structure of nature. Finally, (in Chap. 7) I formulate the conclusions of this inquiry into a general conception of nature, which I term "internal relationalism." According to this view, nature is constituted by its internal temporo-spatial relations. In the course of this investigation, I offer solutions for all the questions and problems I mentioned above, concluding with an answer to the question: "Why is there something rather than nothing?"

To be considered a science, such a research project requires reliable conceptual tools and principles, to insure a systematic and rigorous analysis, and to secure the objectivity of the conclusions. Unfortunately, two and a half millennia of philosophical research have failed to establish generally accepted tools for conceptual analysis. At the beginning of the twentieth century, it was hoped by early analytic philosophers that the development of modern logic would provide the required tools for such an investigation. However, research projects that relied purely on logical tools have failed to provide any great insight into the nature of existence. This is because, notwithstanding the power of modern logic, it was developed to uncover logical connections of inference, rather than to explore connections of meaning between concepts. Currently, there is no generally accepted method for conducting conceptual analysis or effective tools and principles for establishing claims for the existence of conceptual connections.

Hence, the first task in establishing natural philosophy as a science, that is, as a generally accepted and progressing research project, is to develop the conceptual apparatus that is needed for such an inquiry. Although the subjects I explore in this book are very familiar and extensively discussed in the philosophical literature, the inquiry that is undertaken in this book is conducted exclusively within the framework of the conceptual tools and principles of conceptual analysis that are developed here.

My use of the conceptual framework I introduce takes the form of transcendental arguments, in the general sense of attempting to derive the necessary conditions for the meaningfulness of expression in language. However, a common objection to transcendental arguments is that they can only prove that the subject should believe something, not how things actually are (Stroud, 1968, p. 255). Attempting to deduce how things are based on such arguments is thus open to the charge of circularity.

I believe that this objection is valid concerning some transcendental arguments. For example, this is particularly true for transcendental arguments that attempt to confront skepticism by uncovering the necessary conditions for certain types of knowledge. These arguments rely on a complex conceptual structure that the skeptic questions. Furthermore, any conditional statement of the form "if *p* then q," which a transcendental argument might use to substantiate the truth of *p*, presumes *p*'s validity.

However, these objections do not affect my analysis, which ultimately relies on a single concept: existence. The complex structure uncovered in this book is derived from the necessary conditions for the meaningfulness of this concept. In other words, it reveals the essential features of any existence—and, by extension, of nature itself—just as an analysis of the concept of "bachelor" uncovers the necessary conditions for the existence of bachelors. Thus, my analysis is not vulnerable to the criticism that it assumes the applicability or alignment of our concepts with independent reality.

Notwithstanding their ingenuity, many of the existing transcendental arguments in the literature involve complex premises and rely on conceptual principles that lack sufficient justification. Hence, although I shall not ignore the arguments and insights of previous philosophers, I shall limit the discussion of such ideas to those that will be naturally raised in the course of my investigation, and only incorporate arguments and theses that can be coherently formulated and integrated into the conceptual structure that I am founding here. This is mandatory, to ensure the integrity and coherence of this research project and secure the objectivity and evidentness of its conclusions. Nevertheless, the influence of Kant, and later philosophers who followed in his footsteps, in exploring what is sometimes termed "the metaphysics of experience," should be apparent to the reader (these include, inter alia, Bennett (1966); Cassam (1997); Evans (1982); Harrison (1974); Mackie (1980); Stevenson (1982); Strawson (1959, 1966)). Furthermore, my analysis will substantiate many Kantian themes, such as those that relate to time and space as necessary features of nature; the role of the laws of nature in constituting the objective realm, and so on.

I hence begin my research in Chap. 2 by developing the conceptual tools and principles that will guide the subsequent investigation. For this purpose, I analyze the workings of language and introduce a conception of language as a system of coordinates, which is used for mapping the conceptual possibilities that are included in a conceptual scheme. According to this view of language, the meaning of every meaningful component of language is its use in mapping the conceptual space of a conceptual scheme. Based on this conclusion, the principles of conceptual analysis are introduced.

Following the development of the tools for conceptual analysis, I explore in Chap. 3 the meaning of the word "exist." I distinguish between the different uses of "exist" and its inflections. My analysis shows that the concept of existence does not refer to a property. It is rather used to affirm the actualization of one of the conceptual possibilities that are included in a conceptual scheme. This chapter concludes that only what is described by contingent propositions can be meaningfully said to exist. Hence, an exploration of the necessary features of existence must focus on what necessitates the use of contingent propositions for their description.

Chapter 4 begins by exploring the necessary conditions for using contingent propositions in describing what exists. I argue that a contingent proposition is required for describing something if and only if describing it requires this proposition to be both true and false. Hence, I ask the question, how can the same proposition be both true and false without a contradiction? I do not, however, turn to our conceptual scheme to examine how this problem is solved within its confines. This is because our existing conceptual scheme may offer only a possible solution to this problem, but not a necessary one. To expose the necessary features of existence, I hence attempt to construct a conceptual apparatus that allows attributing conflicting truth-values to the same proposition without a contradiction. This conceptual apparatus is then shown to be identical to the temporal indexical expressions in the natural language of past, present, and future. Hence, time, and more specifically the change in the temporal distinctions between past, present, and future—better known as the flow of time—is a necessary element of nature and its most fundamental feature.

There is an ongoing discussion on the nature of indexical expressions in the philosophy of language. Specifically, temporal indexical expressions are at the center of a debate between the A-theory and B-theory of time. Using the conceptual tools developed in Chap. 2, I demonstrate that temporal indexical expressions are unique and indispensable because they allow the same proposition to be expressed in different contexts, enabling it to be both true and false in varying contexts.

Further exploration of the conceptual structure that is involved in giving meaning to the temporal indexical expressions reveals that it must be intricate enough to make the distinction between propositions and what they describe. A conceptual scheme that has the conceptual resources to formulate contingent propositions must be able to acknowledge its own existence in time, the possibility of times in which it did not exist, and the possibility that its descriptions of nature may be wrong. This establishes an objective conception of nature, which is independent of its description by this conceptual scheme. My thus analysis reveals that the distinctions between past, present, and future reflect the fact that nature is described from a temporally located point of view. This raises the question of how these fundamental distinctions can be used to constitute an objective time series—a challenge that will be addressed in Chap. 6.

In Chap. 5, I explore the implication of the conclusion that existence is relative to a point of view that is internal to nature. My main focus in this chapter is to give an account of how a conceptual scheme can form an objective description of nature from a limited point of view in time. I show that this notion of objectivity ultimately relies on the existence of laws of nature, and further uncovers their necessary structure. I begin by making a distinction between inferential and non-inferential descriptions of nature. I show that they both rely on the existence of contingent regularities, which are explored and defined. These are later identified with the laws of nature, which connect occurrences at different times, such that occurrences at one time are both necessary and sufficient conditions for the occurrences at another time. The structure of the laws of nature is further developed and explained, and it is shown that all occurrences in nature must be reciprocally connected by these laws.

In Chap. 6 I define the structure of time and space. At this point, I have already shown how time is (subjectively) constituted as a linear series of occurrences. At this stage of the inquiry, I turn my attention to the constitution of time as an objective feature of nature. I show how, using the conceptual resources uncovered so far, the relations of simultaneity and precedence can be defined; how these relations can be given objective meaning based on the laws of nature; and how distances in time can be measured. Next, I turn my attention to space. I show that it is necessary to give structure to the relations of reciprocal determination between simultaneous existents. Based on this conclusion, I introduce and develop the notion of space, by giving meaning to the idea of locality, and by defining the spatial relations in terms of temporal order of reciprocal determination. Finally, I show that there is a fundamental constant speed of determination between occurrences in space, and further explain how this can be used for measuring space. This chapter concludes that entities in space and time are nothing but reciprocally determining factors, which are themselves constituted by temporo-spatial relations.

Chapter 7 is dedicated to better understanding the conception of nature that is introduced in the previous chapters, that is, the view that I term "internal relationalism." My aim in this chapter is not to further develop the conceptual structure that was introduced in the previous chapters, but rather to explain its meaning and its implications. I begin by returning to the notion of existence, which was first explored in Chap. 3. At this point, the conceptual structure that supports the concept of existence has been revealed, and it implies that existence is a relation. Nature is thus shown to be constituted by pure temporo-spatial relations. I use this conclusion to explain the meaning of the laws of nature and how they constitute time and space.

These laws allow for distinguishing between the different constituents of nature, thus establishing them as separate entities in space while keeping them connected in time. This understanding allows me to explain how nature is constituted. I then return to the notion of time and rely on the insights that I have acquired in the previous discussions in other to explain the arrow of time, that is, the asymmetry between earlier and later occurrences in time, and explore the question of the beginning of nature. Finally, I show how the ancient riddle "why is there something rather than nothing?" is answered by the internal relationalism conception of nature that is developed in this book, by showing that nature necessarily exists.

This book concludes with Chap. 8, in which I summarize the results of my investigation of nature, and outline directions for future research. I argue that empirical science and natural philosophy should not be seen as competing methodologies, but rather as complementary inquiries into nature. Each has its limitations, and hence, for the success of the human endeavor to understand nature, both research projects should be pursued. The end of understanding nature will only be met when the conclusions of both sciences are brought together.

References

Agassi, J. (1969). Leibniz's place in the history of physics. *Journal of the History of Ideas, 30*(3), 331–344.
Bennett, J. (1966). *Kant's analytic*. Cambridge University Press.
Cassam, Q. (1997). *Self and world*. Oxford University Press.
Evans, G. (1982). In J. McDowell (Ed.), *The varieties of reference*. Oxford University Press.
Galilei, G. (2008). *The Essential Galileo* (M. A. Finocchiaro, Trans.). Hackett Publishing.
Gascoigne, J. (2006). The study of nature. In K. Haakonssen (Ed.), *The Cambridge history of eighteenth-century philosophy* (pp. 854–872). Cambridge University Press.
Harrison, R. (1974). *On what there must be*. Oxford University Press.
Hawking, S. W. (1998). *A brief history of time*. Bantam Books.
Hume, D. (1978). In L. A. Selby-Bigge (Ed.), *A treatise of human nature*. Oxford University Press.
Kant, I. (1929). *Critique of pure reason* (N. Kemp-Smith, Trans.). Macmillan.
Kant, I. (2002). Metaphysical foundations of natural science (1786) (M. Friedman, Trans.). In H. Allison & P. Heath (Eds.), *Theoretical philosophy after 1781* (pp. 171–270). Cambridge University Press.
Kerszberg, P. (2006). Natural philosophy. In K. Haakonssen (Ed.), *The Cambridge history of eighteenth-century philosophy* (pp. 873–902). Cambridge University Press.
Kripke, S. A. (1980). *Naming and necessity*. Basil Blackwell.
Mackie, J. L. (1980). *The cement of the universe: A study of causation*. Oxford University Press.
Maxwell, N. (2012). In praise of natural philosophy: A revolution for thought and life. *Philosophia, 40*(4), 705–715.
Smolin, L. (2006). *The trouble with physics*. Houghton Mifflin.
Stevenson, L. (1982). *The metaphysics of experience*. Oxford University Press.
Strawson, P. F. (1959). *Individuals: An essay in descriptive metaphysics*. Routledge.
Strawson, P. F. (1966). *Bounds of sense*. Routledge.
Stroud, B. (1968). Transcendental arguments. *Journal of Philosophy, 65*(9), 241–256.
Unger, R. M., & Smolin, L. (2014). *The singular universe and the reality of time: A proposal in natural philosophy*. Cambridge University Press.
Weinberg, S. (1993). *Dreams of a final theory* (p. 59). Hutchinson.
Yehezkel, G. (2014). *The conceptual structure of reality*. Springer.

Chapter 2
Principles of Conceptual Analysis

Abstract In this chapter, I develop the conceptual tools and principles that will guide the subsequent investigation. For this purpose, I analyze the workings of language and introduce a conception of language as a system of coordinates, which is used for mapping the conceptual possibilities that are included in a conceptual scheme. Based on this conclusion, the principles of conceptual analysis are introduced.

Keywords Conceptual analysis · Conceptual connection · Meaningfulness · System of coordinates

2.1 Introduction

My aim in this chapter is to develop the necessary conceptual tools, and the principles governing their employment, that are needed for the conceptual analysis that is undertaken in the following chapters. The status of natural philosophy as legitimate science hangs on the accuracy of the tools, and objectivity of the principles, by which it is conducted. These insure the rigorousness of the conceptual analysis, and the evidentness of its conclusions.

Conceptual analysis is not only a necessary element of the type of metaphysical inquiry that is involved here. Conceptual analysis is needed in all fields of philosophy, to understand the meaning of the abstract concepts that stand at the center of any philosophical inquiry. Unfortunately, despite millennia of philosophical discussions, a generally accepted conceptual apparatus for conducting such an inquiry has yet to be found. Although the ideas I am advancing throughout this book are greatly influenced by the work of Peter Strawson, his innovative and inspirational book *Individuals* displays the shortcomings of modern conceptual analysis. On the one hand, one finds analyses that rely on a specific, and in many times controversial, theory of meaning, for example, verificationism—as demonstrated in Strawson's analysis of the connection between the concept of self and other minds (Strawson, 1959, Chap. 3). For a similar claim, see Stroud (1968). On other occasions,

G. Yehezkel, *Conceptual Principles of Natural Philosophy*, Synthese Library 495, https://doi.org/10.1007/978-3-031-82108-0_2

conceptual claims seem to ultimately rely on semantic intuitions regarding what one believes is right to say in a thought experiment about a counterfactual scenario. Unfortunately, this leaves us with an unacceptable resolution, in which one must admit that, in the words of Strawson, "I do not think there is any test beyond what we find it satisfactory to say" (Strawson, 1959, p. 81).

Strawson's arguments, particularly those in *Individuals*, have greatly influenced my own reasoning. A tacit principle upon which Strawson often relies, and which will find theoretical justification in the following analysis, is that the meaningfulness of an expression depends on its role in describing the world. Accordingly, Strawson frequently examines the conditions under which an expression or a distinction finds practical application. While I believe this principle to be correct, it currently lacks justification, relying instead on intuition (albeit a sound intuition). Moreover, when applied within the framework of our conceptual scheme, this principle naturally suggests a verificationist theory of meaning; that is, it seems reasonable to presume that language is employed by subjects to describe what exists. This presumption implies a distinction between subject and object, which in turn implies notions of objectivity, and so forth.

Hence, Strawson's arguments are open to criticism, as discussed in the Introduction, that transcendental arguments can only establish that a subject should believe something, or they may risk circularity. For if the existence of the subject is presupposed, it is plausible to conclude that the findings of this analysis merely make explicit what was already implicit within our complex conceptual scheme. Furthermore, if this analysis presumes the existence of a subject utilizing an existing conceptual apparatus, one might question whether this scheme merely reflects subjective beliefs rather than the actual structure of the world. To address these issues, it is necessary to formulate and justify principles of conceptual analysis grounded in a general analysis of language itself. Additionally, both the linguistic analysis and the implementation of these principles in conceptual analysis should avoid presupposing or relying on our established conceptual scheme.Hence, in what follows I develop the principles that will be exclusively used in this book for revealing and establishing the existence of conceptual ties. A conceptual tie is a connection of meaning, which holds between two concepts. Concept x depends on another concept *y* if and only if the meaningfulness of the expression "*x*" depends on the meaningfulness of the expression "*y*." That is, a language that cannot include the concept *y*, is also unable to include the concept *x*. These types of conceptual connections can bind together multiple concepts in complex structures, such that each of the concepts that comprise this construct also depends on it.

The principles of conceptual analysis that I put forward in this chapter are developed and justified based on a model of the workings of language. Fortunately, this endeavor does not necessitate developing a full-blooded theory of meaning, which will, almost inevitably, be controversial. This is not because of a whim on the part of philosophers. Developing a full-blooded theory of meaning necessarily requires explaining the relations between language and the world, which in turn requires developing a metaphysical theory, probably together with a corresponding epistemological account. Unfortunately, developing the latter theory seems to require the very same theory of meaning, and hence involves a vicious circularity.

The ultimate end of any metaphysically-oriented conceptual analysis is to explore the structure of nature. This analysis is supposed to expose the conceptual structure of nature, just as physics attempts to expose the mathematical structure of nature. However, this analysis does not require determining the relations between the world and the language that is used for its description. The analysis of this chapter begins with a study of language as a complex system of meaningful signs. It examines the most general necessary conditions for this system to have meaning, thus enabling it to perform the task of describing the world. The principles of conceptual analysis that I present in this chapter are hence based on a general model of the workings of language. This model relies on the most basic features of language, which I believe are undisputed and should be accepted by any theory of meaning. Based on this model, I extract some necessary conditions for the meaningfulness of expression in language.

I should acknowledge that there is no philosophical idea that is accepted by all. My discussion assumes, for example, that it makes sense to talk about meanings, and even about the meanings of individual words, and not just about the meanings of sentences. This assumption has been criticized before, most notably by Quine (Quine, 1960, Chap. 3; 1970). However, as becomes evident in what follows, my analysis of meaning is minimalist. It does not rely on any type of dubious entity that allegedly constitutes meaning. Furthermore, Quine's criticism is largely based on the lack of what he rightfully sees as the lack of a sufficient account of meaning. Although my analysis does not profess to give a full-blooded theory of meaning, it can be seen as providing an outline of such an account, and should convince the reader that talk about the meanings of words, and conceptual connections, is not just legitimate, but also philosophically worthwhile.

I begin with an analysis of the concept of meaning. As mentioned above, my aim is not to develop a theory of meaning; hence I only advance a limited analysis of the concept of meaning, from which I extract a few undisputed characteristics of the meaning of expressions in a language. In light of these conclusions, I develop a schematic model of the workings of language, according to which language is seen as a system of coordinates, which is used to map the conceptual possibilities that are included in a conceptual scheme. From this model, necessary conditions for the meaningfulness of any component of language are extracted. I then explain the implications of these conditions regarding the conceptual connections between different concepts, thus giving meaning to the idea of a conceptual structure that gives meaning to a component of language. Finally, I summarize the results of my analysis by formulating a model of conceptual analysis, which is utilized in the following chapters.

2.2 (Descriptive) Meaning

In light of the metaphysical motivation of the current investigation, I limit the following analysis of the meaning of expressions in a language to the use of language to describe the world. This limitation does not detract from the importance of the insight, recognized in modern philosophy of language, that language has a variety of uses, and hence that the use of language to describe does not exhaust its meaningful aspects. Nevertheless, I shall note in passing that I do believe that the following analysis and its implications apply to any meaningful aspect of language, although I do not explore this idea in the present book. Hence, in what follows I limit the discussion to the aspect of meaning that might be termed "descriptive meaning." The distinction between descriptive meaning and other types of meaning can be understood in analogy to Frege's distinction between "sense" on the one hand, and "coloring," or "tone" on the other (Frege, 1948, p. 212). For a detailed discussion of this distinction, see Dummett (1981, pp. 2–3, 83–89). Henceforth, for brevity, I shall simply use the term "meaning" to refer to the descriptive aspect of meaning.

To explore the descriptive aspect of meaning, it is necessary to concentrate the attention on declarative sentences, because they are the tool by which language is used to describe. Again, for the sake of brevity, from this point on I simply use the term "sentence" for referring to a declarative sentence.

The term "sentence" refers to a combination of signs that is used to describe the world. It is the meaning of the signs, however, that allows them to describe. In what follows I use the term "proposition" to refer to the meaning of a declarative sentence, that is, the description made by this sentence.

Finally, it is impossible to understand the use of language for describing without relating to the possible success or failure of this use. This description can be right or wrong, and this is reflected in the truth-value of the proposition, which is true if the description is right, and false if the description is wrong. Hence, the meaning of a sentence determines, at least partly, the truth-value of the proposition that is expressed by this sentence. This conclusion, it should be noted, must be accepted by any theory of meaning.

2.3 Meaningful Components of Language

So far, I have concentrated on sentences in a language. However, to understand the meaning of a sentence it is necessary to address the fact that it is (possibly) comprised of more basic meaningful components. This thesis goes back to Frege, who based it on the ability to understand new sentences (Frege, 1984, p. 390). I should emphasize that I do not presume in advance that every sentence must be so comprised. Nevertheless, it is surely a fact that many sentences are comprised of more basic components, and this fact requires an exploration.

I shall therefore turn attention to the relation between a sentence and its meaningful components. I begin by analyzing the contribution of a word to the meaning of a sentence in which it appears, and later expand the discussion to other possible meaningful components of sentences.

Descriptions are made by sentences. Hence, the (descriptive) meaning of a word necessarily relates to the meaning it has as part of a sentence. On the other hand, if the meaning of a word is its contribution to the meaning of sentences, it follows that the meanings of words in a sentence determine, at least partly (possibly together with meaningful components that are not words), the meaning of the sentence. Hence, the meaning of a word, that is, a concept, can be characterized as its contribution to the meaning of sentences in which it occurs. This consideration thus supports Frege's famous claim, "Only in a proposition have the words really a meaning" (Frege, 1980, sec. 60). Finally, it is possible to take into this account the established connection between the meaning of a sentence and the truth-value of the proposition that is expressed by it, that is, the fact that the meaning of a sentence determines, at least partly, the truth-value of the proposition. Hence, it can be concluded that the meaning of a word is its contribution to the determination of the truth-value of the propositions that are expressed by the sentences in which it appears—a characterization that also fits Frege's "sense" (Dummett, 1981, p. 84).

This last conclusion is, I trust, an obvious truism about the connection between the meaning of a word and the truth-value of the proposition that this word helps to express, which any theory of meaning ought to accept. However, as becomes obvious in what follows, much can be extracted from this humble truism, both regarding the workings of language and concerning the necessary conditions for the meaningfulness of expressions in a language.

2.4 The Meaningfulness of Predicates

In this section, I wish to explore the implications of the conclusion of Sect. 2.3, regarding the meaning of a word and its connection with its contribution to the meaning of sentences in which it appears, and thereby its contribution to the determination of the truth-value of the propositions that are expressed by these sentences. What interests me is extracting the necessary conditions for a word to contribute to the meaning of sentences in which it appears, thus acquiring its meaning.

To explore the conditions for the meaningfulness of expressions in a language without any presumption about what constitutes meaning, I suggest viewing language purely as a combination of signs, as it would be viewed by someone who does not understand the language. If one does not know the meaning of the signs in a language, an effective way to explore their meaning, that is, their contribution to the meaning of the sentence in which they appear, would be to examine the effect of a change in these signs. This is not different from an attempt to understand the function of a lever of an unknown machine by attempting to move it and examine the effect. Analogically, the first thing one wants to determine when exploring the meaning of a sign is its function. Since it fulfills its function by appearing in a sentence, to explore its meaning—that is, its contribution to the determination of the truth-value of the proposition that is expressed by that sentence—one should examine the effect of its appearance in that sentence on the determination of the truth-value of the proposition that is expressed by that sentence. This is accomplished by its comparison with the truth-value of the proposition that is expressed when that sign is omitted. Thus, by exploring the effects of all possible legitimate combinations of signs, that is, combinations that express descriptions of the world, the meaning of a sign is explored and exposed.

I start by comparing the meaning of a word as it contributes to the meaning of sentences in which it appears, and more specifically, I begin with a predicate. I aim to uncover the necessary conditions for the meaningfulness of a predicate, that is, the necessary conditions for a predicate to contribute to the meaning of a sentence in which it appears.

Consider the predicate "red." This predicate appears, for example, in sentence (1):

(1) Jack has a red balloon.

The appearance of the predicate "red" in sentence (1) is used to affirm that the property of "being red" is a property of Jack's balloon. This is clear if sentence (1) is compared with sentence (2), in which this predicate is omitted:

(2) Jack has a balloon.

In sentence (2) the predicate red is not ascribed to Jack's balloon. It is a standard convention that if a predicate is not attributed to an object in a sentence, but also not denied, it implies that it is not determined whether the property in question applies to this object or not. There are exceptions to this convention. These, however, do not affect my current analysis.

In passing, I shall mention two reasons for this convention. One reason for this convention is epistemological. That is, because of the idea of objectivity—the possibility that we do not know everything about the world—we need to be able to express a limited description of what exists. Another reason for this convention is that there is no fundamental distinction between affirmation and negation of ordinary predicates. However, I shall not try to explain this claim at the moment. It will be understood once the idea of a conceptual dimension is introduced (in Sect. 2.8). It would then be revealed that deciding whether to denote a property positively or negatively is a matter of arbitrary linguistic convention alone, and each property can be denoted both positively and negatively, using different terms. For example, the color red can be referred to positively as "red," and as "not-der," where "der" is defined as including all colors that are not red.

Turning back to the current analysis, the appearance of the predicate "red" in sentence (1) contributes to the meaning of this sentence, and hence to the determination of the truth-value of the proposition that it expresses. This is demonstrated by the fact that sentence (2) can be true, while sentence (1) is false, that is, in the case in which Jack's balloon is not red.

This difference in the truth-values of the propositions that are expressed by these sentences is explained by the fact that there are two options: either the property of "being red" applies to the balloon, or it does not. The implication of not including the predicate "red" in sentence (2) is refraining from affirming or denying being red as a property of the balloon. Hence, including this predicate in sentence (1), in an affirmative form, implies ascribing the property of "being red" to Jack's balloon, thus contributing to the meaning of the sentence, and hence to the determination of the truth-value of the proposition that is expressed by it.

The contribution of the predicate "red" to the meaning of sentence (1) therefore depends on the possibility of the property "being red" not being a property of Jack's balloon. In other words, the contribution of the predicate "red" to the meaning of sentence (1) in its appearance in the affirmative depends on the possibility that is described by its negation, as expressed in sentence (3):

(3) Jack has a balloon that is not red.

Again, the appearance of the predicate "red" in the negative contributes to the meaning of sentence (3), and hence to the determination of the truth-value of the proposition that is expressed by it. This is again evident by comparing sentence (3) to sentence (2). Sentence (2) can be true while sentence (3) is false—this is the case if Jack's balloon is red. This possibility indicates that the appearance of the predicate "red" in sentence (3) contributes to its meaning.

The conclusion of Sect. 2.3 is that the meaning of a word, that is, a concept, is its contribution to the determination of the truth-value of the propositions that are expressed by the sentences in which it appears. The previous example has shown how the predicate "red" contributes to the determination of the truth-values of the propositions that are expressed by sentences (1) and (3), in which it occurs affirmatively and negatively. This contribution is demonstrated by comparing the possible truth-values of these propositions with the possible truth-value of the proposition that is expressed by sentence (2), in which the predicate is not included. Thus, if Jack has a blue balloon, sentence (2) expresses a true proposition, while adding the predicate "red" in the affirmative form results in sentence (1), which expresses a false proposition. Similarly, if Jack has a red balloon, while sentence (2) expresses a true proposition, adding the predicate "red" in the negative form results in sentence (3), which expresses a false proposition. This is a straightforward demonstration of how the appearance of the predicate "red" in a sentence contributes to the determination of the truth-value of the proposition that is expressed by it, thus reflecting its contribution to the meaning of this sentence—a contribution that constitutes its meaning.

I should emphasize that the above analysis does not rely on any presumption regarding what constitutes the meaning of a sentence, and more specifically, it does not rely on the premise that the meaning of a sentence is its truth conditions. I have only examined the effect of the appearance of the predicate in a sentence on the truth-value of the proposition that is expressed by it as a means for determining a possible change in the meaning of this sentence. Nor does this analysis rely on an assumption about what constitutes the meaning of a word, for example, the assumption that the meaning of a word is its reference. Furthermore, to uncover the effect of the appearance of a predicate in a sentence, I have examined a case in which one word is added to a sentence. Because my analysis did not involve two words with the same reference but a different "sense" (in Fregean terminology), the (alleged) distinction between sense and reference does not influence this analysis.

The conclusion of my analysis should therefore be accepted by any theory of meaning. It implies that if the occurrence of a word in a sentence cannot possibly have any effect on the determination of the truth-value of the proposition that is expressed, then this word does not contribute to the meaning of this sentence. The effect on the determination of the truth-value of the proposition is examined by comparison to the truth-value of the proposition that is expressed when the word is omitted, and this effect is used to determine the contribution of the word to the meaning of the sentence that expresses this proposition.

This analysis implies that if balloons were necessarily red, then the occurrence of the predicate "red" would have not contributed to the meaning of sentence (1). Admittedly, I rely in this context on the idea of a conceptual necessity, that is, a necessity that is based on meaning (rather than, for example, the type of necessity suggested by Kripke (Kripke, 1980, pp. 38–39). However, there is nothing mysterious or dubious about this type of necessity. I refer to the possibility that the term "balloon" would have been defined not simply as an "inflated bag," but rather as an "inflated red bag." Hence, if this were the case, the claim that a balloon is red would be a conceptual truth, just like the claim that a widower was married.

The fact that if balloons were necessarily red the predicate "red" would have not contributed to the meaning of sentence (1) is evident once its truth-value is compared with the truth-value of sentence (2), which does not include this predicate. If balloons would have necessarily been red, sentence (1) and sentence (2) would have expressed logically equivalent propositions, that is, their truth-values would be necessarily identical. If sentence (2) expresses a true proposition, so does sentence (1); and if sentence (2) expresses a false proposition, so does sentence (1). This is a clear indication that adding the predicate "red" to sentence (2), which results in sentence (1), does not change the proposition that is expressed. That is, the predicate "red" does not contribute to the meaning of sentence (1), for it does not contribute to the determination of the truth-value of the proposition it expresses.

An example may help in understanding the claim I am making here. Consider, for example, the sentence "Jack is a widower who has been married." Since part of the definition of a widower is a man who has been married, the predicate "has been married" does not contribute to the meaning of this sentence. This is evident from the fact that this sentence expresses a proposition that is logically equivalent to the proposition that is expressed by the sentence "Jack is a widower." For it shows that adding the predicate "has been married" does not contribute to the determination of the truth-value of the proposition that is expressed by the resulting sentence. In other words, nothing new is learned, if the fact that he has been married is added to the description that Jack is a widower, because it is a conceptual truth that a widower has been married.

My analysis focused on the occurrence of the predicate "red" in the affirmative, that is, when it is used to ascribe the property of being red to Jack's balloon. The conclusion of my analysis, however, is general and equally applies to the negative occurrences of the predicate "red," as in sentence (3), that is, in instances in which it is used to deny that Jack's balloon is red. The contribution of the negative appearance of the predicate "red" in sentence (3) is explained by the possibility of the balloon being red. If it was not conceptually possible for balloons to be red, the occurrence of the predicate red in sentence (3) would have not contributed to the meaning of this sentence. This is demonstrated by the fact that its appearance in the negative could have not contributed to the determination of the truth-value of the proposition that is expressed by sentence (3). Again, this is reflected by the fact that in this hypothetical scenario sentences (2) and (3) would have expressed logically equivalent propositions.

The conclusion that emerges from this analysis is that a predicate can contribute to the meaning of a sentence only if both its appearance in the affirmative and in the negative formulate possible descriptions of the world. If this is not the case, then adding the predicate to the sentence, in the affirmative or in the negative, either results in what is conceptually necessary or in a contradiction. In the former case, the appearance of this predicate adds to the determination of the truth-value of the proposition that is expressed because it is conceptually necessary (as exemplified in the sentence, "Jack is a widower who has been married"). In the latter case, its appearance does not contribute to the meaning of the sentence because it results in a contradiction, that is, it results in something that cannot be used to describe anything, and hence has no (descriptive) meaning (as exemplified in the sentence, "Jack is a widower who has not been married").

I should emphasize, that the claim that a contradiction, that is, a sentence that is unable to describe anything, has no (descriptive) meaning does not imply that it is complete nonsense. Indeed, it should be reminded that the negation of something nonsensical is itself nonsensical, and that negation of a contradiction is a tautology, that is, a necessary truth. It is hence advisable to follow Wittgenstein's terminology in the *Tractatus Logico-Philosophicus,* and describe contradictions and tautologies as senseless, but not nonsensical (Wittgenstein, 1963, para. 4.461). In this context, it is interesting to note that it is possible to explain Wittgenstein's views on contradictions and tautologies with the help of similar considerations to those presented here (Carruthers, 1989, p. 63). I hence suggest viewing tautologies as empty propositions (in terms of descriptive meaning), which can be used to reflect conceptual connections, which constitute our conceptual scheme.

I have relied, in the previous analysis, on the possibility of omitting a predicate from a sentence and compared the truth-values of the propositions that are expressed by the two sentences, to examine the contribution of the predicate to the determination of the truth-value of the proposition that is expressed the sentence in which it appears. For this purpose, I have chosen to analyze sentences in which omitting the predicate nevertheless results in a declarative sentence. The conclusions of my analysis are not affected by the fact that there are sentences in which this omission is only possible if the sentence is rephrased. This is the case, for example, with the sentence "Jack's balloon is red," in which omitting the predicate "red" results in "Jack's balloon." The original sentence, however, made two claims: that Jack has a balloon, and that it is red. Hence, omitting the claim regarding the color of the balloon leaves us with the claim that "Jack has a balloon," with which it is possible to compare the truth-value of the proposition that is expressed by the sentence "Jack's balloon is red."

It is hence important to note that the possibility, or impossibility, of omitting a predicate from a sentence does not affect the generality of my conclusion. The possibility of omitting a predicate from a sentence is not a necessary condition for it to contribute to the meaning of that sentence. Rather, it is only a means I use in my analysis, whose purpose is to expose the contribution of a predicate to a sentence, and the necessary conditions for its meaningful appearance in a sentence.

My analysis shows that a predicate can contribute to the meaning of a sentence in which it appears only if the property it ascribes is a contingent property of the thing to which it is ascribed. Thus, the connection between the meaning of a predicate, that is, its contribution to the meaning of sentences in which it appears, and the possibilities of its appearance in a sentence is as follows: a predicate contributes to the meaning of a sentence, in the affirmative or in the negative, only if its appearance in both cases formulate sentences that express contingent propositions. If this is not the case, then in one of these options (of appearances in the affirmative or in the negative), it does not contribute to the meaning of the sentence since it results in a contradiction, and hence fails to describe anything; while in the other option it reiterates what is already expressed by the sentence, that is, it indicates what is necessary, and hence fails to contribute to its meaning. Hence, *the meaning of a predicate is its ability, by its occurrence in the affirmative or in the negative in a sentence, to identify one of two conceptual possibilities.*

2.5 Language as a System of Coordinates

The conclusion of the previous analysis can be generalized, and formulated with regard not only to any word in a sentence, but to any meaningful component of a sentence. The meaning of every meaningful component in a language is its contribution to the meaning of the sentences in which it appears. For a component of a sentence to contribute to its meaning, its contribution to the meaning of this sentence should not indicate what is either conceptually necessary or what is conceptually impossible. That is, it must be possible to both affirm and deny any alleged contribution, of a possible meaningful component, to the meaning of a sentence. The meaning of every meaningful component of a sentence is therefore its ability, by its occurrence in a sentence, to determine which option, of at least two distinct conceptual possibilities, obtains.

The implications of this conclusion can be demonstrated by an analogy to a coordinate system on a map. If a conceptual scheme is understood as a collection of conceptual possibilities, then it is likened to a regular map, which includes possible locations of a traveler. Language is equated, in this analogy, to a system of coordinates on a map. It is a system of signs that is used for a methodical mapping of all the possibilities that are included in a conceptual scheme. A sentence, as a structured collection of signs, is a combination of coordinates that enables denoting a specific conceptual possibility. Furthermore, a declarative sentence represents an attempt to denote an actualized possibility, just as a combination of coordinates on a map can be used to designate the actual location of a traveler (in contrast to a designated target, for example).

The meaning of every meaningful component in a language is determined by its role in this systematic mapping of the conceptual possibilities that are included in a conceptual scheme. Its meaning is its contribution to the determination of the conceptual possibility that is designated by the sentences in which it appears.

Although all the meaningful components in a language acquire their meaning from their role in this systematic mapping, not all the meaningful components of a language serve the same function. Predicates, for example, function as coordinates for a specific conceptual dimension of the conceptual space of possibilities that are included in our conceptual scheme, similar to the different coordinates that are used for each of the two dimensions of a map. Each predicate is used to designate (at least) one of the possibilities that are included in this dimension. There are, as I shall demonstrate in what follows, meaningful components that have other uses, although they also acquire their meaning from their role in this system of coordinates.

There is no minimum number of coordinates that is necessary to formulate a sentence. A sentence can be more detailed, that is, include coordinates of multiple conceptual dimensions, and thus help to pinpoint a more specific point in the conceptual space of possibilities, for example, "Jack has a red balloon;" or it may include fewer coordinates, and thus be more general, for example, "Jack has a balloon," or even "Jack has something."

As mentioned above, every meaningful component of language acquires its meaning from its use in this system of coordinates, although not all these components function as coordinates that designate a certain conceptual possibility on a specific conceptual dimension. For example, negation, as exemplified by the word "not," does not function as a coordinate of a specific conceptual dimension, but rather acquires its meaning from its function in this systematic mapping while it is coupled with words that are used as coordinates on a specific conceptual dimension. If a word designates one of the possibilities included in a certain conceptual dimension, when it is coupled with a negation word, the phrase designates the other possibilities (one or more) on this conceptual dimension. In the previous example, while the word "red" denotes one of the possible colors of Jack's balloon, when coupled with a negation word, as exemplified by the sentence, "Jack's balloon is not red," the resulting phrase denotes the other possible colors of the balloon, that is, the colors that are not red.

In light of these insights, it now becomes clear why, for the word "red" to contribute to the meaning of the sentence "Jack's balloon is red," its occurrence both in the affirmative and in the negative must describe a conceptual possibility. It is because the word "red" is used as a coordinate on the conceptual dimension that

Table 2.1 The conceptual dimension of the predicate "Greek"

Socrates is Greek
Socrates is not Greek

includes all the possible colors of Jack's balloon. These possibilities can generally be divided into two, the color red, and the colors that are not red. If there were no other possibilities other than Jack's balloon being red, the word "red" would not have been needed as a coordinate on the dimension of colors, for there would have been no possibilities to choose from. If this was the case, the word "red" would not have contributed to the meaning of this sentence, because it would serve no function as a coordinate.

A minimal conceptual dimension must therefore include at least two distinct conceptual possibilities. If a conceptual dimension includes only two possibilities, it can be mapped with the help of a single word. This word can denote either of the two possibilities, while the other possibility would be designated by the same word coupled with a negation word. Thus, if black and white were the only possible colors, it would have been possible, although not necessary, to use "black" and "not-black" to differentiate between these possibilities. If a conceptual dimension includes more than two possibilities, more words are required to map all the different possibilities that are included in this dimension, as is the case with the dimension of colors, which includes "red," "green," "blue," and so on.

I use the expression "conceptual space" to describe the totality of the conceptual dimensions in which a proposition is located. To uncover the conceptual space in which a certain sentence is located, it is necessary to reveal the conceptual dimensions in which every meaningful component of this sentence is located. For this purpose, it is first and foremost advisable to examine the possibility of adding negations to the words that comprise this sentence. For example, one of the conceptual dimensions that are included in the sentence "Socrates is Greek" is demonstrated in Table 2.1.

The phrase "not Greek" includes many different possibilities, which can be further explicated. However, I shall not explore this further. Instead, I wish to turn my attention to other components of sentences. So far, I have only discussed predicates. However, this sentence is located in a conceptual space that includes more dimensions than that of the predicate "Greek." More specifically, I wish to explore the dimension of the subject of this sentence.

2.6 The Conceptual Dimension of a Subject

It is now the time to turn the attention to another conceptual component of sentences, that is, the subject, and explore the necessary conditions for it to contribute to the meaning of a sentence. In what follows I concentrate on names since they present the biggest challenge for the application of the type of analysis presented here to the subject of a sentence.

To begin with, notice that names are words that contribute to the determination of the truth-value of the propositions that are expressed by the sentences in which these names appear. Hence, names are meaningful signs in language, according to my use of the term "meaning." Furthermore, because names are meaningful signs, it follows that the conclusions of my previous analysis, regarding the necessary conditions for the meaningfulness of signs in a language, apply to them as they apply to any meaningful component of a sentence.

For a name to contribute to the meaning of a sentence in which it appears, it must be conceptually possible to deny what it adds to the meaning of the sentence. Otherwise, as I have already demonstrated in the case of predicates, its alleged contribution would add nothing to the description that the sentence makes. That is, its appearance in a sentence would not contribute to the determination of the truth-value of the proposition that is expressed by this sentence.

However, in trying to apply to names the conclusions of the previous analysis, regarding the necessary conditions for a predicate to contribute to the meaning of a sentence in which it appears, a problem arises. While it is clear what is meant by negating what a predicate adds to a description, that is, denying a property as an attribute of something, it is not clear what can be negated in the case of a name that appears as a subject—obviously an object cannot be negated.

However, recall that the previous analysis concludes that it should be conceptually possible to negate what a predicate contributes to the meaning of a sentence, rather than the property itself. Similarly, it is not the object that is referred to by a name that it should be possible to negate, as a necessary condition for the meaningfulness of the appearance of this name in a sentence. It is rather what this name contributes to the meaning of the sentence, that is, what it contributes to the formulation of the description made by the sentence. Hence, applying the conclusion of the previous analysis to different meaningful components of a sentence is not always a uniform and straightforward exercise. It can take different forms, depending on the various ways in which different meaningful components of a sentence contribute to its meaning.

The key to applying the conclusions of the previous analysis to names is therefore to identify the specific contribution of names to the descriptions that are expressed by the sentences in which they appear. Turning the attention to the sentence "Socrates is Greek," to determine the contribution of the name "Socrates" to this sentence, it is helpful to ask what the name adds to the description. This can be examined by imagining a case of someone who fails to hear the beginning of the sentence, and hence knows that there is someone Greek, but not who is Greek. In answer to the question "Who is Greek?" it would be possible to answer, "It is Socrates," and this answer would add to the incomplete description that was heard. It is also possible to answer, "It is *not Socrates*," and this demonstrated how the contribution of a name to the meaning of a sentence can be negated.

Hence, the name "Socrates" contributes to the meaning of the sentence "Socrates is Greek" because of the conceptual possibility of different subjects, other than Socrates, being Greek. The name "Socrates" contributes to the meaning of the sentence by indicating which of the possible subjects is referred to. If it was only conceptually possible for Socrates to be Greek, the name "Socrates" would have not contributed to the meaning of this sentence.

This conclusion coheres with what became known as the Generality Principle, first introduced by Strawson (Strawson, 1959, p. 99). However, I should point out that the consideration upon which Strawson relies to justify this principle is different from the considerations I have presented here. Strawson's line of thinking refers to an analysis of the concept of a predicate, rather than the role of the subject in a sentence. It is based on the claim, which is possibly influenced by Kant's distinction between concepts and intuitions, that "the idea of a predicate is correlative with that of a range of distinguishable individuals of which the predicate can be significantly [...] affirmed" (Strawson, 1959, p. 99). Furthermore, notwithstanding a superficial similarity, my claim should be carefully distinguished from Evans' Generality Constraint, because Evans' principle is a constraint on the explanation of the ability to *understand* an existing conceptual structure of a language, rather than a constraint on the meaningfulness of this structure, as I argue here (Evans, 1982, pp. 100–105).

My analysis concludes that the same necessary conditions for contributing to the meaning of sentences in which they appear apply to both predicates and names. In both cases, the contribution they make to the meaning of a sentence is by indicating which of (at least) two conceptual possibilities is referred to. Hence, in both cases, their meaning depends on the existence of a dimension of possibilities. In the case of a predicate, it is a dimension of properties that it is conceptually possible to ascribe to an object. In the case of names, it is a dimension of objects to which it is conceptually possible to, for example, ascribe a predicate. Thus, the contribution of the name "Socrates" to the meaning of the sentence "Socrates is Greek" depends on the possibility of other subjects, other than Socrates, being Greek. The meaningfulness of the appearance of the word "Socrates" in this sentence arises from its use for indicating which, out of all those to which the predicate "Greek" can meaningfully be ascribed, is being referred to. The conceptual possibilities on the dimension on which "Socrates" is used as a coordinate can again be generally divided into two possibilities, that is, that Socrates is Greek or that Someone other than Socrates is Greek.

The sentence "Socrates is Greek" is therefore located in a conceptual space that includes at least two different dimensions, each for the words "Socrates" and "Greek." These words are used as coordinates on these dimensions and can appear in this sentence alone or together with a negation word. The conceptual space in which this sentence is located is, therefore, (partially) described in Table 2.2.

Notice that it is possible to simplify the notation, thereby also highlighting the analogy between names and predicates, by the convention that adding the negation "not" to a name *a*, that is, "not-*a*," implies that someone other than *a* is referred to.

Table 2.2 The conceptual space of the sentence "Socrates is Greek"

Socrates is Greek	Someone that is not Socrates is Greek
Socrates is not Greek	Someone that is not Socrates is not Greek

However, my intention in the present discussion is not to formulate a perfect language, or even to suggest an improvement to the notation of natural language. It is merely to understand the necessary conditions for the meaningfulness of expressions in a language.

According to some views, the simple grammatical structure of the sentence "Socrates is Greek" actually conceals a conceptually complex structure. Thus, according to Russell's theory of descriptions, the underlying structure of this sentence is, for example, "The philosopher who drank the poison is Greek." Furthermore, the definite description "The philosopher who drank the poison," in turn, can itself be analyzed into more basic components, that reveal the complex logical structure that constitutes it. However, I should emphasize that such claims do not affect the conclusions of my analysis, because my analysis applies to any meaningful component of language, whether simple or complex.

The previous analysis shows that the necessary conditions for contributing to the meaning of sentences apply equally to both names and predicates. That is, in both cases, affirming and denying that contribution should describe a conceptual possibility. This does not imply that there are no differences between a subject and a predicate, including differences in the implication of their denial. For example, the sentences "Socrates is Greek" and "Socrates is not Greek" express inconsistent propositions; while the sentences "Socrates is Greek" and "not-Socrates is Greek" express consistent propositions. However, these differences have no bearing on the conclusions of the previous analysis.

So far, I have discussed the necessary conditions for the meaningfulness of subjects and predicates. However, it might be argued that I have ignored the predicative element. This element, continues the objection, is represented even in simple propositions, such as "Socrates is Greek," with the word "is." According to this suggestion, while the name "Socrates" indicates one of the possible subjects to which a predicate can be meaningfully ascribed; the predicate "Greek" indicates one of the possible properties that can conceptually be ascribed to the subject; and the third element is the predicative element, which determines whether the property is affirmed or denied of the subject.

Notice that this suggestion undermines the conclusion of the previous analysis, by suggesting that the contribution of a word to the meaning of a sentence does not depend on the possibility of its denial, but rather on the possibility of substituting it with another word, which may have no logical or conceptual connection with the original word, thus affecting the meaning of the sentence. Thus, according to this suggestion, the word "red" contributes to the meaning of the sentence (1) "Jack has a red balloon" because it designates one of the different predicates, such as "big", that can be conceptually ascribed to a balloon.

To begin with the last implication of this objection, it should be noted that the possibility of replacing "red" with "big" is not a sufficient condition for the word "red" to contribute to the meaning of sentences. As I have argued, if balloons can be only red, the appearance of the word "red" does not contribute to the meaning of sentence (1), as evidenced by the possibility of omitting it without affecting the truth-value of the proposition that is expressed. The possibility of ascribing other predicates to a balloon does not affect the meaningfulness of the appearance of the predicate "red" in this sentence. The same applies to the word "Socrates," in the sentence "Socrates is Greek." If it was only conceptually possible for Socrates to be Greek, the word "Socrates" would have not contributed to the meaning of this sentence, notwithstanding the possibility of replacing it with "He," (while referring to Socrates), "Someone," or "The philosopher who drank the poison."

Furthermore, it is impossible to distinguish between the predicates and the predicative elements in sentences. This is evident from the fact that the conditions for the meaningfulness of the predicate and the alleged predicative element are the same. They both involve the conceptual possibilities of both affirming and denying the predicate. It follows that there is no predicative element, distinct from the predicate, that is a conceptual component of a sentence.

Admittedly, the word "is" does contribute to the meaning of sentences in which it appears. This is clearer in the sentence, "Jack is married," for example. For, in this case, the temporal implication of the use of this word becomes apparent. To expose its contribution to the meaning of the sentence, it should be compared with other tenses of this auxiliary verb, that is "was," and "will be." Hence, the word "is" is a conceptual component in some sentences in which it appears, and is used as a coordinate on a temporal dimension that includes the distinction between past, present, and future.

2.7 Other Conceptual Components of Sentences

So far, I have discussed the contribution of words to the meaning of sentences. However, there are other conceptual components of a sentence, that is, elements that contribute to the meaning of a sentence. In some cases, for example, the order of the words in a sentence contributes to its meaning. For example, in the sentence "Jack loves Jill" the order of the names contributes to the meaning of the sentence, and therefore the conclusions of the previous analysis, regarding the conditions for a component of a sentence to contribute to its meaning, equally applies to the order of the names.

To uncover the contribution of the order of the names to the meaning of the sentence "Jack loves Jill," it is helpful to think of a case in which the order of the names is forgotten. In this case, it is remembered that there is a love relationship between Jack and Jill, but it is forgotten who loves who. The order of the names tells us that it is Jack that loves Jill. The order of the names contributes to the meaning of the sentence because of a different possibility, that is, the possibility that Jill loves Jack. Again, it must be possible to deny what the order of the names contributes to the meaning of the sentence, although it is more complicated in this case. This can be expressed by the claim that "there is a love relationship between Jack and Jill, but it is not the case that Jack loves Jill." The existence of the love relationship between Jack and Jill leaves more than one possibility open, and the order of the names is used to identify one of these possibilities. Again, if it were not for the existence of at least two mutually exclusive possibilities, that is, that Jack loves Jill and that Jack does not love Jill, the order of the names in this sentence would have not contributed to the meaning of this sentence.

As should be clear by now, the conceptual components of a sentence are varied. They can include signs, the order of the signs, or any other feature of a sentence. For example, it is possible to adopt a notation in which the order of the words in a sentence is replaced by italics. Thus, according to this notation, the sentences "*Jack* loves Jill" and "Jill loves *Jack*" would have the same meaning, which according to the prevailing notation is expressed by the sentence "Jack loves Jill." In this alternative notation, the order of the names will not have any significance.

Again, it is shown that the conclusion of the previous analysis applies to any conceptual component of a sentence, even if it is not a sign. That is, for a component of a sentence to contribute to the meaning of this sentence, the contribution it makes to the meaning of this sentence could not be conceptually trivial or result in a contradiction. That is, it should be possible to both affirm and deny what this component contributes to the description made by this sentence without a contradiction.

2.8 Conceptual Dimension

In previous sections, my analysis focused on the meaningfulness of components of a sentence. I have concentrated my attention on revealing the necessary conditions for a component in language to contribute to the meaning of sentences. In this section, I wish to turn the attention to a broader question, which concerns the necessary conditions for something to be a conceptual component in a language in general.

The conditions for the meaningfulness of components in a language are intrinsically tied with the conditions for their meaningful appearance as a component in a sentence. The reason is that the notion of meaning that is under discussion here is a descriptive meaning, and descriptions are performed by the use of sentences. However, the question I currently ask refers to the conceptual scheme in general, which is understood to include all the possibilities that enable the formulation of meaningful sentences in a language.

The previous analysis shows that the meaningful components in a language acquire their meaning in mapping the conceptual space of possibilities that are included in a conceptual scheme. This space of possibilities is not amorphic. It includes different conceptual dimensions that comprise different possibilities. Some of the meaningful components in language acquire their meaning by their use as coordinates on these dimensions, most simply by referring to one of the possibilities that are included in such a dimension.

In what follows I further explore the idea of a conceptual dimension. I focus the discussion on the conceptual dimension of predicates, because of the complexity of these dimensions, which requires special attention.

As I have previously argued, a minimal conceptual dimension includes only two possibilities. This is the minimal number of possibilities because if there is only one possibility there is no dimension that needs to be mapped. In other words, if there is only one possibility, it is ipso facto a necessary property, and hence ascribing it to anything would fail to contribute to the description that is made.

In the simplest case, therefore, a dimension includes two conceptual possibilities. As mentioned previously, although it can be mapped with the use of two different predicates, it is also possible to map this dimension with the help of only one predicate, for example, *P*, that indicates one of these possibilities, while the other possibility will be referred to by its denial, that is, not-*P*. It follows from this that every predicate in language can be seen as a coordinate on a separate conceptual dimension, which includes itself and its denial. This view, however, of the structure of a conceptual scheme would inevitably lead to overlooking a variety of important conceptual ties that obtain between different predicates.

There are indeed simple cases in which a conceptual dimension includes only two possibilities, that can be mapped with the help of a predicate *P* and its negation. However, the shortcomings of this strategy, of interpreting every predicate as located on a separate dimension, can be seen if it is assumed that, as is many times the case in natural language, a different predicate *F* is used as a name for the possibility not-*P*. If *P* is construed as located on a different conceptual dimension than *F*, the conceptual connection that obtains between these predicates will be overlooked.

In more complex cases, a conceptual dimension includes more than two possibilities. An example of such a conceptual dimension is found in the variety of predicates that are used for describing the marital status. These include such predicates as, for example, single, married, divorced, and widow\widower. The reason for including all these conceptual possibilities on a single conceptual dimension is to reflect the conceptual connections that obtain between them, such as, for example, that being a bachelor excludes the possibility of being married. Hence, the possibility of being married is included in the general possibility (which includes several distinct possibilities) of being a not-bachelor. Thus, demonstrating the conceptual connections that obtain between these predicates requires locating them on a single conceptual dimension.

Notice, however, that the predicates that are located on a single conceptual dimension do not necessarily exclude one another. For example, the predicate "available" can be used to denote the possibilities of being single, divorced, or widowed. What is necessary is that for each meaningful predicate *P*, on any conceptual dimension, there is a possibility (or possibilities) that is (are) excluded by this predicate, that is, the possibility that is denoted by not-*P*. For, as the analysis of the previous section has shown that if negating a predicate does not denote a conceptual possibility, this predicate cannot contribute to the meaning of sentences, and is hence rendered meaningless.

A conceptual dimension can therefore be a complex structure. However, at the most fundamental level, it must include at least two conflicting possibilities. Predicates that find their use, that is, their meaning, as coordinates on this dimension either designate one of these fundamental possibilities or are a disjunction of these basic possibilities. Predicates can also be defined by a combination of possibilities of different conceptual dimensions. This introduces the possibility of various conceptual connections between different dimensions. Hence, the first task of any conceptual analysis is to determine what are the fundamental conceptual dimensions and explore the connections that obtain between different dimensions.

So far, I have discussed the conditions for a predicate to contribute to the meaning of a sentence in which it appears. The fact that the appearance of a predicate in a sentence fails to contribute to the meaning of that sentence does not imply that this predicate is meaningless. For this predicate may successfully contribute to the meaning of other sentences. However, it now becomes clear that there are necessary conditions for a predicate to have meaning in a language in general. If not for the conceptual structure of a language, which reflects the conceptual scheme it maps, that allows the formulation of sentences in which a predicate contributes to their meaning, it would indeed be meaningless.

There are two main ways in which a predicate can lose its meaning altogether. The first is if the conceptual dimension in which it is located is lost; the second is if there are no objects to which it can be meaningfully ascribed (see Sect. 2.9). In both cases, this predicate would fail to contribute to the meaning of the sentences in which it appears.

The importance of determining the predicates that are located on a specific conceptual dimension, therefore, relates to the conceptual connection of dependence that obtains between them. If the meaning of a predicate is its role as a coordinate on a specific conceptual dimension, then the existence of more than one possibility on this dimension is a necessary condition for the meaningfulness of this predicate. In the case of a dimension that includes only two possibilities, the dependence is straightforward, and the two predicates that are used to map this dimension are interdependent. In the case that there are more than two possibilities, the interdependence between the predicates on this dimension is obviously weaker but can nevertheless be important for exploring conceptual connections.

2.9 Conceptual Basis of a Predicate

As mentioned earlier, the existence of a conceptual dimension on which a predicate *P* is located is just one necessary condition for the meaningfulness of the predicate *P*. This requires the conceptual scheme to include the possibility of not-*P*. However, this is not the only necessary condition for the meaningfulness of *P*. For example, as the analysis of Sect. 2.3 shows, ascribing a predicate *P* to an object *a* is only meaningful if it is conceptually possible for *a* to be both *P* and not-*P*. Thus, for example, if it is conceptually impossible for the number 5 to be married, it is meaningless to claim that it is married and it is meaningless to claim that it is not married. If only human beings can have a marital status, then the meaningfulness of the predicate "married" in a language depends on the inclusion of the category of human beings in that language.

Hence, the meaningfulness of a predicate *P* in a language depends on the category of objects to which this predicate can be meaningfully ascribed—what can be termed the conceptual basis of predicate *P*. A conceptual scheme that does not include the conceptual basis of predicate *P* would not include predicate *P* as well.

Exploring and defining the category of objects to which a predicate can be meaningfully ascribed is therefore important for identifying another necessary condition for the meaningfulness of this predicate, thus achieving a better understanding of its meaning. This category of objects must include all and only all those objects to which this predicate can be meaningfully ascribed. The more accurately this category is determined, the more accurate the conditions for the meaningfulness of this predicate are identified.

The dependence of a predicate *P* on the category of objects to which it can be meaningfully ascribed is a type of conceptual connection that obtains between two predicates, that is, between two concepts. This is because a category of objects is defined with the help of a predicate, for example, *Q*. Thus, it is a species of a conceptual connection of dependence, such that a conceptual scheme that does not include the category of objects *Q*, would not include predicate *P* either, nor could this conceptual scheme include the conceptual dimension on which predicate *P* is located.

2.10 Conceptual Analysis

The model of language as a system of coordination is introduced here to extract principles for conceptual analysis. Due to this focus, I have not attempted to develop this model into a comprehensive theory of meaning. Consequently, the reader may find that some questions remain unanswered. For instance, for clarity, I have only discussed simple declarative sentences. However, one might question how the concept of language as a coordination system applies to more complex sentence types, such as relational sentences, or to other non-descriptive uses of language, including unarticulated linguistic constituents.

While I believe that the model of language as a system of coordination could potentially serve as a foundation for a comprehensive theory of meaning, pursuing such a project far exceeds the scope of this book. I do note, however, that the principles I have formulated here apply equally to any declarative sentence. For example, relational terms can be analyzed as predicates with two or more positions. The principles for analyzing these sentences are therefore the same, with the exception that, as I have shown in Sect. 2.7, word order can be a conceptual component of these sentences. Similarly, the general principles underlying language function are applicable to any declarative sentence, regardless of complexity.

Analyzing other uses of language, however, requires a more complex approach. Since these uses are not central to the current project, I will limit my comments to noting that any meaningful linguistic component can be analyzed by considering the implications of its presence or existence within a speech act, specifically with regard to the intended goal of that act. This can be examined by comparison to speech acts where the component is either omitted or replaced. For example, descriptive meaning can be analyzed based on its contribution to determining the truth-value of a proposition expressed by a sentence. Similarly, any other meaningful linguistic component, whether articulated or unarticulated, can be assessed by the effect of its presence or existence on the intended outcome of the speech act, whether that is to convey emotion, direct action, or achieve other communicative purposes.

Returning to the main topic, I will now outline the conclusions of this analysis of language with respect to the method of conceptual analysis. The meaning of any meaningful component in a language is determined by its use in sentences, that is, its contribution to the meaning of sentences in which it appears. Thus, the ability to include a meaningful component in a language depends on the complexity, or richness, of this language. If a language is not rich enough to allow formulating sentences in which this component can be meaningfully used, this component cannot be included in this language. By exposing the necessary structure that enables giving meaning to such a component, for example, a concept *P*, the conceptual connections that obtain between the concept *P* and other concepts in a language are revealed, that is, the conceptual structure that supports the concept *P* is uncovered. This is the end of any conceptual analysis.

The structure of a language can be seen as a representation of a conceptual scheme, or a conceptual space of possibilities. This view of language does not carry any ontological commitments. The idea of a conceptual scheme is used as a representation of the structure that is included in a certain language, which allows explaining how components in this language acquire their meaning. According to this view of language, any meaningful component in a language acquires its meaning in its use as part of a system of coordinates, which maps the conceptual space of its conceptual scheme.

Traditionally, exposing conceptual connections was thought to be exhausted by logical connections. However, logical connections represent only one type of conceptual connection. The analysis of this chapter shows that there are other types of conceptual connections, such as the conceptual connections of dependence that obtains between concepts that are located on the same conceptual dimension, and the connections between predicates and the category of objects to which they can be meaningfully ascribed.

When the connection between two concepts is examined, that is, when it is inquired whether concept *x* depends upon concept *y*, what is actually in question is whether a language that does not include concept *y* can include concept *x*. To examine this in the confines of our conceptual scheme, what must be done is to narrow the conceptual space by eliminating the conceptual possibility that is denoted by concept *y*, and examine whether the word "*x*" is useful for mapping the remaining conceptual space. If the answer is negative, then concept *x* depends on concept *y*.

The exploration of the conceptual connections that obtain between different concepts in our conceptual scheme is what Strawson calls descriptive metaphysics (Strawson, 1959, p. 9). However, it should be noted that the term "descriptive" is somewhat misleading, for conceptual connections are necessary connections. Nevertheless, this type of exploration always leaves open the following concern.

Suppose that it is found that, in our conceptual scheme, time is a necessary feature of the world, that is, that the concept of existence requires temporal distinctions. The question nevertheless remains, whether time is a necessary feature of the world. For the reader may wonder whether it is possible to replace the temporal distinctions with other distinctions that would fulfill a similar function, thus enabling us to support the concept of existence even without temporal distinctions.

In the next chapter, I explore the concept of existence and the necessary conditions for its inclusion in a conceptual scheme. In the later chapters, I examine how these necessary conditions can be fulfilled, thus exposing the necessary features of anything that exists. My analysis of how the necessary conditions for the concept of existence can be fulfilled does not rely, however, on our conceptual scheme. I instead attempt to construct a language from nothing and expose the conceptual structure that is necessary for including the concept of existence in any conceptual scheme. This method of inquiry will ensure that its conclusions expose the necessary features of existence.

References

Carruthers, P. (1989). *Tractarian semantics*. Basil Blackwell.

Dummett, M. (1981). *Frege: Philosophy of language*. Harvard University Press.

Evans, G. (1982). In J. McDowell (Ed.), *The varieties of reference*. Oxford University Press.

Frege, G. (1948). Sense and reference. *The Philosophical Review, 57*(3), 209–230.

Frege, G. (1980). *The foundations of arithmetic* (J. L. Austin, Trans.). Northwestern University Press.

Frege, G. (1984). Compound thoughts. In B. F. McGuinness (Ed.), & P. Geach & R. H. Stoothoff (Trans.), *Collected papers on mathematics, logic and philosophy* (pp. 390–406). Basil Blackwell.

Kripke, S. A. (1980). *Naming and necessity*. Basil Blackwell.

Quine, W. V. O. (1960). *Word and object*. MIT Press.

Quine, W. V. O. (1970). On the reasons of indeterminacy of translation. *The Journal of Philosophy, 67*, 178–183.

Strawson, P. F. (1959). *Individuals: An essay in descriptive metaphysics*. Routledge.

Stroud, B. (1968). Transcendental arguments. *Journal of Philosophy, 65*(9), 241–256.

Wittgenstein, L. (1963). *Tractatus logico-philosophicus* (D. P. Pears & B. F. McGuinness, Trans.). Routledge & Kegan Paul.

Chapter 3
The Meaning of Existence

Abstract In this chapter, I analyze the use of the word "exist" and its inflections in the confines of the conception of language as a system of coordinates. My analysis shows that the concept of existence does not refer to an ordinary property, but rather affirms the actualization of one of the conceptual possibilities that are included in a conceptual scheme. This chapter concludes that only what is described by contingent propositions can be meaningfully said to exist.

Keywords Conceptual scheme · Conceptual structure · Contingency · Existence

3.1 Introduction

To uncover the necessary features of existence, one must first understand the meaning of the term "exist." In this chapter, I explore the use of this term, to uncover the function it plays in mapping the possibilities that are included in our conceptual scheme. The question is whether the distinction between existence and non-existence is located on a specific conceptual dimension, or whether this distinction plays a different role in the systematic mapping of our conceptual scheme.

The question I ask is not entirely divorced from the use of the concept of existence in logic. However, the traditional analysis of the concept of existence was undertaken in terms of a different view of the workings of language than the view that is directing the present investigation, that is, the view of language as a system of coordinates. Hence, to ensure that it is unbiased, my investigation is undertaken by disregarding the prevalent view of this concept. Nevertheless, a comparison of the results of my investigation with the interpretation of the concept of existence in modern logic will find that these conceptions of existence, although formulated in different terms, do cohere.

Furthermore, modern logic's purpose is somewhat limited compared to the large scope of the current analysis. Logic is limited to the analysis of the conceptual connections that are relevant for uncovering logical relations, that is, relations of inference, and their main motivation is creating the tools for the assessment of such

G. Yehezkel, *Conceptual Principles of Natural Philosophy*, Synthese Library 495, https://doi.org/10.1007/978-3-031-82108-0_3

relations. The current analysis investigates conceptual connections of dependence in general. Its purpose is metaphysical, that is, to uncover conceptual structures that support certain concepts—the concept of existence in this specific analysis. Notice that the two investigations differ in scope. First, connections of inference relate to propositions, while the present investigation focuses on concepts. Admittedly, my analysis of meaning relies on the premise that a word can acquire meaning only through its use in sentences, which are used to express propositions. However, I believe that logic's focus on propositions limits its ability to observe more direct connections between concepts, specifically connections of dependence. A logical connection between two propositions can indicate a conceptual connection between the concepts they contain, such that the meaningfulness of one concept depends on the other. However, the reverse does not hold: a conceptual connection of dependence between two concepts is not necessarily demonstrated in a logical connection between propositions that include these concepts. Thus, the type of conceptual analysis I employ here is broader in scope, as it directly addresses relationships between concepts, including relations of dependence that are not necessarily logical.I begin my analysis by examining the use of existence in terms of the model of the working of language I introduced in Chap. 2. My conclusion coheres with the current prevalent understanding of existence, as demonstrated in modern logic, according to which existence is not an ordinary predicate, that is, it is not used as a coordinate on a specific conceptual dimension. The word "exist" does not add any coordinate to the sentence in which it appears—it is not a property of objects—but is rather used to affirm that the conceptual possibility that is denoted by the coordinates that are included in this sentence is actualized.

Finally, I explore the implications of my analysis of the concept of existence. I show that any meaningful language, that is, language that enables the formulation of contingent propositions, has the conceptual resources to support the notion of existence. In other words, only things that require the use of contingent propositions for their description can be meaningfully said to exist. Exploring the necessary conditions of existence is therefore identical to exploring the necessary conditions for being described by contingent propositions.

3.2 The Uses of "Exist" and Its Inflections in Natural Language

According to the conclusion of Chap. 2, the meaning of any meaningful component in a language is its contribution to the meaning of sentences in which it appears. Hence, to understand the meaning of "existence," it is necessary to reveal the function it serves in terms of the view of the workings of the language as a system of coordinates. For this purpose, I turn my attention to the uses of the word "exist" and its inflections in natural language, that is, its contribution to the meaning of sentences in which it appears.

As with any other meaningful component in language, the word "exist" contributes to the meaning of a sentence in which it occurs only if both possibilities of its occurrence in this sentence, that is, in the affirmative and in the negative, describe a conceptual possibility. That is, the word "exist," in the sentence "φ exists," can contribute to the meaning of this sentence only if both this sentence and the sentence "φ does not exist" express a conceptual possibility, that is, a contingent proposition.

Notwithstanding the question of whether existence is a property of objects or not, it follows from the conditions for the meaningfulness of expressions in language that the word "exist" acquires its meaning, that is, finds its use in language, by being ascribed to things whose existence is contingent. In other words, the conceptual basis of the concept of existence is the category of things whose existence is contingent.

I use the term "things," while describing the conceptual basis of the concept of existence, because the word "exist" and its inflections are ascribed to a variety of different logical and conceptual categories. In sentences of the structure "*a* exists" it is conjoined with a name, for example "Moses existed;" in sentences of the structure "*F*'s exist" (or "$\exists xFx$" in predicate calculus) it is conjoined with a predicate, for example, "Centaurs exist;" and in sentences of the structure "the existing state of affairs is that *p*" it is conjoined with complete sentences, for example, "the existing state of affairs is that 'a few wealthy persons control the economy'."

3.3 "Exist" as a Sentential Connective

A satisfactory account of the meaning of the word "exist" and its inflections must provide a uniform and coherent account of all of its uses. I begin by examining the use of the word exist as a sentential connective, that is, when it is conjoined with a complete sentence. This is because the use of existence as a sentential connective provides a simple way to examine the contribution of the word "existing" to the meaning of the sentence, that is, to the determination of the truth-value of the proposition that is expressed by the resulting sentence. This can be easily done by comparing the truth-value of the resulting compound proposition to the truth-value of the basic proposition, to which the term "existing" is conjoined.

As the previous example shows, the word "existing" is used as a sentential connective as part of the expression "the existing state of affairs is that…" It might be objected that the term "existing" determines only part of the meaning of this expression, and hence it is impossible to explore the meaning of the term "existing" simply by exploring the effect of this sentential connective on the determination of the truth-value of the proposition that is expressed by the compound sentence. However, the expression "The…state of affairs is that…," which is used to conjoin the notion of existence with a complete sentence, is merely used as a grammatical element of this sentence. It hence fails to contribute to the meaning of the resulting compound sentence. This is evident from the impossibility of negating its alleged contribution to the meaning of the resulting sentence, which is a necessary condition for its meaningful appearance in this sentence.

Although the term "The…state of affairs is that…" is used to connect the word "existing" with a sentence *p*, it is therefore merely a "grammatical connector," whose use is to create a grammatically well-formed sentence. The conceptual structure of the sentence "The existing state of affairs is that *p*" is, therefore, "p exists." In this respect, existence is similar to the negation word "not." For, as a sentential connective, the word "not" appears as part of the grammatically compound expression "It is not true that…" However, it is a simple truth-functional sentential connective, which in logic is represented by the logical structure "not-p." In both instances, that of "exist" and "not," a grammatical expression is used to conjoin a simple one-word (and one place) sentential connective with a sentence.

For simplicity, I ignore in what follows the inflections of the word "exist" and the grammatical connective, which are of no conceptual importance, and simply use the connective "…exist." This sentential connective is admittedly a technical expression, but in this respect, it is similar to the truth-functional sentential connective "not…," which also does not reflect the grammatical use of negation in natural language.

It might be objected that, notwithstanding its grammatical appearance, the word "existing," in the sentence "The existing state of affairs is that *p*," does not function as a sentential connective. According to this suggestion, exposing the logical structure of this sentence would reveal that "existing" actually functions as a quantifier, that is, a second-order predicate, as it is interpreted in modern logic.

In answer to this objection, it should first be noted that even if this claim is true, in the context of the present analysis it cannot be presumed that this analysis of existence is true. As previously explained, this view of the function of the term "existing" was not developed in the confines of the conception of language that is the backdrop of the current investigation. Furthermore, the possibility of reformulating sentences in which existence seemingly functions as a sentential connective in such a way that it conforms with the standard view of existence, does not imply that existence cannot be also used as a sentential connective. If two formulations of the same sentence are possible, this fact alone cannot determine which formulation, if either, is more basic, and exposes the "true structure" of this sentence. Finally, I should emphasize again that the seeming use of existence as a sentential connective is of specific value. It allows us easily to examine the contribution of the term "exist" to the meaning of the sentence in which it occurs, and this consideration is unaffected by the question of whether it is indeed a sentential connective or not.

As a sentential connective, the word "exist" is used as a truth-functional sentential connective, that is, the truth-value of the proposition that is expressed by "*p* exists" is determined by the truth-value of *p*. The compound proposition is true if and only if the proposition *p* is true, and false if and only if the proposition *p* is false.

Since existence, as a sentential connective, can be conjoined with any meaningful sentence in a language, its conceptual basis is the whole language. That is, if it is possible to formulate a meaningful sentence, it is possible to find a use for existence as a sentential connective.

However, notwithstanding the general applicability of existence as a sentential connective, it seems that it fails to contribute to the meaning of the sentences in which it appears. This is evident from the fact that it fails to contribute to the determination of the truth-value of the propositions that are expressed by these sentences. It is completely superfluous, since "*p* exists" claims nothing more than simply *p*—the truth-value of both propositions is always the same, and hence adding existence as a sentential connective does not contribute to the determination of the truth-value of the proposition that is expressed by the conjoined sentence.

I should make clear that my conclusion does not rely on the assumption that the meaning of a sentence is its truth conditions. That is, I am not relying on the premise that if any two sentences express logically equivalent propositions, they are ipso fact synonymous. I am arguing that if adding an expression to a sentence fails to change its truth conditions, this expression fails to contribute to the meaning of the resulting sentence. This follows from the characterization of the meaning of an expression as its contribution to the determination of the truth-value of the propositions that are expressed by the sentences in which it appears.

What is strange, however, is that both the sentence "*p* exists" and the sentence "*p* does not exist" express contingent propositions. Thus, for example, both "The state of affairs that 'a few wealthy persons control the economy' exists," and "The state of affairs that 'a few wealthy persons control the economy' does not exist," express contingent propositions. That is, it is possible to negate the contribution of the meaning of the word "exist" to the meaning of these sentences without contradiction. Furthermore, negating this contribution results in a change in the truth-value of the proposition that is expressed, which shows that the appearance of the word "exist" is not meaningless. Nevertheless, it does not seem to reiterate any specific distinction that is denoted in the sentence with which is it conjoined. This is clear both from the fact that this analysis applies to any meaningful sentence *p* to which it is conjoined, and from the fact that negating existence in these sentences does not result in a contradiction.

It can therefore be concluded that there are two distinct conceptual possibilities, which the word "exist" is used to denote, by its appearance in the affirmative or in the negative. Its peculiar function is similar, in this respect, to the sentential connective "It is true that…," or simply "true...," as used in logic. The sentences "It is true that *p*" and "*p*" express logically equivalent propositions. Nevertheless, both "It is true that *p*" and "It is not true that *p*" express contingent propositions (assuming that *p* is a contingent proposition). This indicates that the sentential connective "true" is not vacuous.

Part of the explanation for the unique behavior of "true" and "exist" as sentential connectives is that in these cases a standard convention in language does not apply. According to this standard convention, if an expression is not included in a sentence, it is neither affirmed nor denied. In other words, not including an expression is a way of refraining from determining whether it applies or not. Thus, for example, the sentence "Jack has a balloon" does not imply anything about the color of the balloon. Hence, adding "red" to the sentence contributes to the determination of the possible state of affairs that is described by this sentence, and hence to the determination of the truth-value of the proposition that is expressed by it.

However, in the case of "exist" and "true," in their use as sentential connectives, the implication of not including these expressions in a sentence is their affirmation. Thus, it follows from *p* both that "The existing state of affairs is that *p*," and that "It is true that *p*."

The explanation for the uniqueness of these terms, and their resemblance, is found in the essence of a declarative sentence. A declarative sentence is a description of what *exists*, and expresses a *true* proposition if and only if it successfully describes what *exists*. This is correct even in the case of negative existentials, for example, "No dragons exist," as is evident from the possibility of formulating this sentence as "The existing state of affairs is that no dragons exist." This is the explanation for why *p* is logically equivalent to both "*p* exists" and "It is true that *p*." It is not an arbitrary convention (Wittgenstein, 1963, para. 4.062). Rather, the ideas of (descriptive) meaning, truth, and existence are intrinsically connected.

The fact that "*p* exists" does not say more than "*p*", as implied by their logical equivalence, indicates that the word "exist" and its inflections do not add any distinction to the meaning of the sentence in which they appear. They do not function as a coordinate on a specific conceptual dimension, which helps to designate a more specific conceptual possibility than that which is described by the expression to which they are conjoined. In its use as a sentential connective, the word "exist" affirms (or denies if it appears in the negative) that the conceptual possibility, described by the sentence to which it is conjoined, is actualized. In terms of the conception of language as a system of coordinates, the distinction between existence and non-existence is not located on a separate dimension in the conceptual space of possibilities. It simply relates to any possibility that is included in this conceptual space and is used to affirm or deny its actualization.

3.4 Other Uses of "Exist"

It is now time to show how the explanation of the use of existence as a sentential connective can be extended to explain its other uses. As the previous analysis shows, when existence is conjoined with a sentence, for example. "'*a* is *F*' exists," it is used to reaffirm the actualization of the conceptual possibility that is described by the sentence to which it is conjoined. In this example, this conceptual possibility is determined with the help of (at least) two separate coordinates, that is, the words "*a*" and "*F*," which constitute this sentence.

As already mentioned, a conceptual possibility can be described more generally or more specifically. This is achieved by omitting coordinates or including more coordinates from other conceptual dimensions. For example, the conceptual possibility that is described by the sentence "Moses led the Israelites out of Egypt" can be described more generally by omitting some of the coordinates it includes, as in the sentence "Moses led the Israelites;" or more specifically, by adding more coordinates, as in the sentence "Moses led the Israelites out of Egypt 3,500 years ago."

Keeping this in mind, we can now explain the use of the concept of existence in sentences such as "*a* exists" and "*F* exist." This explanation is best undertaken by extending the understanding of "exist" as a sentential connective to its other applications.

Let us begin with an example of the sentential connective use of "exist" in the proposition "the existing state of affairs is that 'Moses is leading the Israelites'." Here, "exist" is used to affirm the realization of a complex state of affairs that includes both the individual referred to as "Moses" and the predicate "leading the Israelites." We can generalize this by omitting one coordinate of the complex predicate, resulting in "the existing state of affairs is that 'Moses is leading'."

Suppose we omit the predicate altogether. This can be done by refraining from specifying any particular property ascribed to Moses in the original proposition. In this case, we are left with an affirmation of only one coordinate. This could be expressed, somewhat awkwardly, as "the existing state of affairs is that 'Moses is something'." This formulation might be less awkward if we imagine someone who did not hear the end of the sentence but reports what they did hear. However, language provides us with a more elegant way to express this idea. Since we only intend to affirm the realization of one coordinate, we can simply say, "Moses exists." Similarly, if we omit the component "Moses," the resulting proposition would be, "There exists someone who is leading the Israelites."

In conclusion, while in the sentence "'*a* is *F*' exists," existence is used to reaffirm the actualization of the conceptual possibility that is denoted by the two coordinates, "*a*" and "*F*," in the sentences "*a* exists" and "*F* exist" existence is used to affirm the actualization of more general conceptual possibilities, that is, the possibilities that are denoted by only one of the two coordinates "*a*" and "*F*." Thus, the sentence "*a* exists" affirms that the conceptual possibility, that is, the possible entity, that is denoted by "*a*," is actualized without ascribing to it any property; while the sentence "*F* exist" affirms the actualization of the conceptual possibility, that is, the property, that is denoted by "*F*," without ascribing it to any entity.

It is now time to compare the conclusion of my analysis with other prominent modern views of existence—namely, those of Kant, Frege, and Russell. Of these influential thinkers, Kant's view is closest to my analysis (Kant, 1929, p. B626, 1979, pp. 57–61). Kant distinguishes existence from ordinary predicates and argues that the concept of existence serves not as a real predicate but merely as a logical one. Indeed, Kant seems to echo my claim that the concept of existence relates to the conceptual possibility designated by the other coordinates with which it is conjoined when he writes that:

> …nothing can have been added to the concept, which expresses merely what is possible, by my thinking its object (through the expression 'it is') as given absolutely. Otherwise stated, the real contains no more than the merely possible. (Kant, 1929, p. B627)

Kant argues that existence is a predicate of a concept, rather than of the objects that fall under this concept. In this sense, Kant sees existence not as a first-level predicate but as a second-level one. However, Kant acknowledges not only general concepts but also concepts of individuals or specific objects. The primary difference, then, is that the conception of existence I introduce here is more general, as it also accounts for the use of "exist" as a sentential connective.

Frege expresses a similar view, arguing that existence is a second-level predicate—a property of concepts rather than of the objects subsumed by them (Frege, 1970, pp. 48–50, 1980, pp. 64–65). However, despite the similar formulation, a fundamental difference exists between these views. Kant's notion of concepts allows him to explain propositions like "a exists" as ascriptions of existence to a concept of an individual. Frege, however, draws a strict distinction between concepts and objects as two types of entities, each referred to by different types of expressions. As a result, Frege dismisses the concept of existence in propositions of the type "a exists," making his view unable to account for this function of the concept of existence.

In Russell's framework, existence is again treated as a second-order property, specifically a property of propositional functions, which aligns with my analysis. For Russell, the proposition "Moses exists" asserts that a propositional function of the type "*x* is *F*" (where *F* is a definite description of Moses) holds true for at least one value of *x* (Russell, 1956, pp. 242–243). A propositional function is an expression with an undetermined constituent, such that "*x* is *F*" results from omitting the "determined constituent" (or, in my terms, "the coordinate") *a* from the proposition "*a* is *F*." This interpretation parallels my view that, in a sentence of the type "*F* exists," existence affirms that the conceptual possibility determined by *F* is actualized. However, this view, too, does not account for all uses of the concept of existence.

3.5 The Univocity of Existence

It may be objected that I have not sufficiently proven that existence does not function as a property, that is, that it does not function as a coordinate on a specific conceptual dimension. I believe this is clear in the case of the use of existence as a sentential connective, because in this use it is evident that the concept of existence does not add any distinction to the description of the conceptual possibility that is denoted by the sentence with which it is conjoined. However, it might be argued that I assume, without sufficient proof, that the word "exist" and its inflections serve the same function in all its uses, including when it is conjoined with parts of sentences.

To begin with, it should be agreed that the use of existence in both "*a* exists" and "*F*'s exist" is the same, as is evident from the validity of the inference: Stromboli exists and is an island volcano, so an island volcano exists (Mackie, 1976, p. 257). The question is therefore whether this use is the same as its use as a sentential connective. To answer this question, it is sufficient to show that its use as a sentential connective is identical with one of its other two uses since it was shown that the latter uses are identical.

To examine the use of existence in sentences of the type "*a* exists," the conceptual structure of this sentence ought to be compared with a simple sentence in which a property is ascribed to an object, that is, "*a* is *F*." The conceptual structure of the sentence "*a* is *F*" can be examined by negating the predicate in this sentence. Thus, the sentence "*a* is not-*F*" ascribes to the object *a* one of the other properties that are found in the conceptual dimension in which *F* is located. Notice that there is a fundamental difference between negating the predicate and negating the whole sentence. For, when the whole sentence is negated, that is, "It is not true that '*a* is *F*'," it does not imply that any possibility is actualized.

The reason for the difference between negating the predicate *F* and negating the whole sentence is obvious. The sentence "*a* is *F*" is a combination of (at least) two coordinates that designates one of the many possibilities that are included in our conceptual scheme. If one of the coordinates in this sentence is negated, the resulting sentence denotes another conceptual possibility in our conceptual scheme. However, if the whole sentence is negated, the resulting sentence does not indicate any actualized conceptual possibility, but rather denies the actualization of the conceptual possibility that is denoted by the original sentence. Thus, when a property is ascribed to something in a sentence, there are two different distinctions in this sentence that can be negated separately, as well as the whole sentence.

Sentences of the type "*a* exists," on the other hand, have a different conceptual structure, as revealed once the two words that comprise this sentence are negated. If indeed existence is a property, this sentence should be located in a conceptual space that includes the possibilities that are described in Table 3.1.

If existence is a property, that is, used as a coordinate on a specific conceptual dimension, sentence (3.1.2) should be claiming the actualization of a conceptual possibility. However, sentence (3.1.2) is synonymous with a negation of sentence (3.1.1), that is, "It is not true that '*a* exists'." The fact that denying existence is not different from denying the whole sentence indicates that this sentence includes only one coordinate, that is, "*a*," and despite its grammatical appearance, the term "exists" does not ascribe any property to *a*. It relates to the conceptual possibility that is identified by the word "a," that is, a possible entity, and affirms its actualization.

It can therefore be concluded, that in sentences of the type "*a* exists" the word "exists" serves the same function as in its use as a sentential connective. This suffices to show that "exist" is univocal across all its uses. This conclusion highlights the advantages of analyzing the concept of existence through the conception of language as a system of coordinates, both in relation to the view that existence is an ordinary first-order predicate and the view that it is a second-order predicate, as interpreted in modern logic. Only within the framework of language as a system of coordinates—where existence is used to affirm the actualization of a conceptual

Table 3.1 The conceptual space of the sentence "a exists"

(3.1.1) *a* exists	(3.1.2) *a* does not exist
(3.1.3) Not-*a* exists	(3.1.4) Not-*a* does not exist

possibility, rather than to specify a position on a conceptual dimension—can we consistently explain all its uses in language.

3.6 What Exists

The analysis of the concept of existence shows that there is a close connection between existence and (descriptive) meaning. Existential sentences of all types are used to affirm the actualization of one of the conceptual possibilities that are included in a conceptual scheme. The connection between existence and meaning is reflected by the fact that any affirmative declarative sentence is likewise used to affirm that a conceptual possibility is actualized.

The fact that existence is used to affirm the actualization of a conceptual possibility, similarly to a descriptive sentence, implies that the conditions for the meaningfulness of the word "exist" are identical with the conditions for the meaningfulness of expressions in language in general. In other words, every meaningful language has the conceptual resources to include the concept of existence.

As the analysis of Chap. 2 shows, the meaning of every conceptual component in a language is its contribution to the meaning of declarative sentences in which it appears. This contribution depends on the use of this component to denote a conceptual possibility, either directly or when conjoined with other meaningful components of the language. Indeed, the same is true of complete sentences, which themselves are meaningful components of language, and can contribute to the meaning of compound sentences in which they appear as components. Affirmative sentences not only denote a certain conceptual possibility but also affirm the actualization of this possibility.

The conclusion of the analysis of the concept of existence is that an affirmative existential statement is a claim for the actualization of a conceptual possibility. Hence, if a conceptual scheme is understood as the totality of all conceptual possibilities that a language enables us to describe, it can be concluded that every conceptual scheme can find a use for the notion of existence, or in other words, every meaningful language has the conceptual resources to include the concept of existence. This is because while the meaning of most of the words in a language is their use in the mapping of a specific conceptual dimension, existence can be used to affirm the actualization of any possibility that is included in a conceptual scheme.

The necessary and sufficient conditions for the meaningfulness of "exist" and its inflections are therefore identical to the necessary and sufficient conditions for meaningfulness in general. These, however, are identical to the conditions for the ability to formulate contingent propositions in language. This is because meaningful sentences express contingent propositions. This is clear from the fact that the conditions for the meaningfulness of expressions in language, according to which its occurrence, both in the affirmative and in the negative, ought to describe a conceptual possibility, equally apply to sentences. Hence, a meaningful sentence is a sentence that expresses a contingent proposition. Any meaningful language has the conceptual resources to express contingent propositions.

It can therefore be concluded that the necessary and sufficient condition for the meaningfulness of the word "exist" is the conceptual apparatus that enables the formulation of contingent propositions. This connection between existence and contingency is expected, given the fact that existence is used to affirm the actualization of conceptual possibilities, which are themselves described by contingent propositions.

Only things that are described by the use of contingent propositions can therefore be meaningfully said to exist. This implies that logical and mathematical truths do not describe an ontological realm of mathematical and logical entities. Only entities whose existence is contingent can be meaningfully said to exist. The totality of these contingent beings and state of affairs is what we call "nature."

To understand what constitutes existence, the question that must be answered is, what features require the use of contingent propositions for their description? For describing any existent requires the use of contingent propositions. The features that require for their description the use of contingent propositions are the necessary features of nature. Although it may sound paradoxical, the goal of the following inquiry is therefore to uncover the necessary structure of contingent existence.

References

Frege, G. (1970). On concept and object. In P. Geach & M. Black (Eds.), *Translations from the philosophical writings of Gottlob Frege* (pp. 42–55). Basil Blackwell.

Frege, G. (1980). *The foundations of arithmetic* (J. L. Austin, Trans.). Northwestern University Press.

Kant, I. (1929). *Critique of pure reason* (N. Kemp-Smith, Trans.). Macmillan.

Kant, I. (1979). *The one possible basis for a demonstration of the existence of God* (G. Treash, Trans.). University of Nebraska Press.

Mackie, J. L. (1976). The riddle of existence. *Aristotelian Society Supplementary, 50*, 247–266.

Russell, B. (1956). The philosophy of logical atomism. In *Logic and knowledge*. George Allen & Unwin.

Wittgenstein, L. (1963). *Tractatus logico philosophicus* (D. P. Pears & B. F. McGuinness, Trans.). Routledge & Kegan Paul.

Chapter 4
The Temporal Foundation of Nature

Abstract In this chapter, I explore the necessary conditions for using contingent propositions for describing nature. For this purpose, I construct the conceptual apparatus that allows attributing conflicting truth-values to the same proposition without a contradiction. This conceptual apparatus is found to be identical to the temporal indexical expressions of past, present, and future. Hence, time, and more specifically the "flow of time," is found to be a necessary and the most fundamental feature of nature. I further explore this conceptual structure and show that it relies on a notion of objectivity, which distinguishes between propositions and what they describe.

Keywords Contingency · Flow of time · Indexicals · Objectivity · Time

4.1 Introduction

The conclusion of Chap. 3 is that the only things that could be meaningfully said to exist are those that are described by contingent propositions. To uncover the necessary features of nature, the question that guides this inquiry is: what features require the use of contingent propositions for their description?

In this chapter, I argue that only in one case is there a need for contingent propositions for describing something—in the case in which the description requires the use of propositions that are both true and false. This appears to be a contradiction. Hence, the focus of my analysis is exploring how a language can describe what exists using propositions that are both true and false without a contradiction.

My analysis does not rely on our conceptual scheme in the course of this investigation. That is, I do not attempt to examine how our conceptual scheme enables us to formulate contingent propositions. This limitation is required to ensure that the conclusion of this investigation indeed exposes the necessary conditions of existence, rather than contingent properties that merely characterize the actual world, according to our conceptual scheme.

G. Yehezkel, *Conceptual Principles of Natural Philosophy*, Synthese Library 495, https://doi.org/10.1007/978-3-031-82108-0_4

For the purpose of this investigation, I attempt to formulate a language from nothing and develop the conceptual apparatus that is required to formulate contingent propositions. That is, I attempt to build, from its foundation, the conceptual apparatus that enables the same propositions to be both true and false without a contradiction. So far, my analysis has only shaped the tools and prepared the ground for the real task of this book, that is, the task of rebuilding nature from nothing. Thus, the formulation of this new language mirrors the creation of nature, and the conceptual structures of this language describe the necessary features of nature.

The attempt to construct a language in which contingent propositions can be formulated is extremely abstract and includes abstruse and subtle considerations that are not easy to follow. In this respect, this is the most challenging chapter in this book, but it is also the most important. Russell writes that abstract studies, such as philosophical logic, are so difficult and elusive that a really good philosopher is the one who manages to think about it once in 6 months for half a minute (Russell, 1956, p. 185). The rest of the time, suggests Russell, we are thinking about symbols because they are tangible. However, the most difficult thing to do, as shall be evident in this chapter, is to think of symbols abstractly, without conferring any meaning on them. Indeed, I have tried to carry out this investigation without any reliance on our conceptual scheme. This is the only way, I believe, to understand the nature of time and its significance as a necessary feature of nature. However, the decision to conduct this investigation in a completely abstract manner is also responsible for its more challenging aspects.

For a reader more interested in the conclusions of this investigation, and less in the intricate analysis that leads to it, it is not necessary to delve into the subtleties of the analysis of Sects. 4.3, 4.4, 4.5, 4.6, 4.7, 4.8, and 4.9, in which I develop the conceptual construct that supports the contingency of propositions in language. The important aspect of this analysis is the final structure, which is found to be (in Sect. 4.10) identical to the temporal indexical expressions in our conceptual apparatus, which therefore explains the contingency of propositions in language. What is remarkable in this conclusion is the fact that the construction of this apparatus, which is undertaken in complete disregard for our existing language, is found to be perfectly identical to our conception of time. This proves that all things necessarily exist in time.

Thus, the conclusion of my analysis in this chapter is that time, and more specifically the distinctions between past, present, and future, are the explanation for the need for contingent propositions for describing what exists. Hence, the change in the temporal distinctions between past, present, and future—better known as the flow of time—is a necessary feature of nature and its most fundamental aspect. Furthermore, to explain how a proposition can be both true and false, a conceptual scheme must include the conceptual resources that allow it to acknowledge the existence of its propositions, as well as its existence as part of nature. This implies that this conceptual scheme must be able to acknowledge the possibility of times in which it did not exist, as well as the possibility that its descriptions of nature are mistaken. Thus, the use of contingent propositions for describing nature is shown to rely on an objective conception of existence, which is demonstrated in a distinction between what exists on the one hand and its description by that very same conceptual scheme on the other. This conceptual structure is the basis for the complex structure that will be developed in the following chapters.

4.2 The Use of Contingent Propositions

The question that guides this inquiry is: What are the features of nature that require, for their description, the use of contingent propositions? To better understand this question, recall that, according to the conception of language as a system of coordinates, the meaning of every meaningful component in a language is its use in sentences that describe what exists.

The problem that I am facing now can be explained thus. A language allows some combinations of signs. These are the conceptual structures that have a use in describing what exists, and hence are meaningful. Some formulations of signs are not recognized in a language as syntactically well-formed sentences at all, for example, "The good is more or less identical than the beautiful" (Wittgenstein, 1963, para. 4.003). However, even amongst the well-formed sentences, some sentences are identified by the rules of the language as failed descriptions of what exists, that is, contradictions. Since contradictions cannot describe what exists, they are not meaningful. The negation of a contradiction is a tautology, or conceptual truth which, due to being a negation of a contradiction, is likewise not meaningful. As I argued in Chap. 2, although it is not nonsensical, it is nevertheless empty, and thus fails to describe what exists. The only category of meaningful sentences in a language is the category of sentences that express contingent propositions, that is, propositions that can be possibly true and possibly false. The question is, why are such sentences needed to describe what exists?

The difficulty of answering this question can be demonstrated by the following consideration. Take any sentence, for example, "*a* is *F*." It is either true or false. If it expresses a false proposition, it does not describe what actually exists, and hence it is not needed for describing what exists. Hence, it seems that it is possible to describe what exists with the help of a language in which the sentence "*a* is *F*" is identified as expressing a contradiction, and its negation as expressing a conceptual truth. On the other hand, if "*a* is *F*" does describe what exists, that is, expresses a true proposition, then its negation is useless for describing what exists. Hence, it seems possible to describe what exists with the help of a language in which "*a* is *F*" is identified as expressing a conceptual truth, and its negation as expressing a contradiction. Thus, it seems that to describe what exists, there is no need at all for a language that enables the formulation of sentences that express contingent propositions. However, the analysis of Chap. 3 has shown that this conclusion is false since only contingent propositions can be meaningfully said to describe what exists.

It might be suggested that the need for contingent propositions is merely a constraint on the richness of the language, rather than a constraint on what exists. That is, it might be thought that all that is required is that a language would include more conceptual possibilities than those that are required for true descriptions of what exists. Thus, suppose that Jack's balloon is red. One might argue that although in this case there is nothing that necessitates the use of language that includes the conceptual possibility of Jack's balloon not being red, there is nothing that prohibits the use of a language that includes such excessive structure, that is, a language that includes the possibility of Jack's balloon being not red. Surely, the objection continues, a problem arises only if one is attempting to describe what exists with the help of a language that includes insufficient conceptual possibilities. There is no difficulty in describing something with the help of a language with excessive conceptual structure. All that is required, to explain the contingency of propositions in a language, is that nature would allow, rather than necessitates, the use of contingent propositions for its description.

If indeed this objection is valid, it would undermine the current inquiry. For there seems to be nothing to *prohibit* the use of excessive conceptual structure even when it is unnecessary for describing what exists. However, recall that the meaning of every meaningful component in a language is *its use* for describing what exists, and more specifically, its contribution to the meaning of sentences that describe what exists. Hence, an apparatus that has no use in describing what exists cannot be a meaningful part of a language. Only what has a use, that is, a need, in describing what exists can be meaningful. If only contingent propositions can be meaningfully said to describe what exists, nature must include features that *necessitate* the use of contingent propositions for its description.

Another suggestion for explaining the need for contingent propositions for describing what exists, without resorting to special characteristics of nature, requires turning to epistemology. According to this suggestion, although there is no need for false propositions for describing what exists, it is impossible to formulate a language in which such propositions are contradictions. For this would require a priori knowledge of what exists. Because it is not known what actually exists, there is a need for a language that includes contingent propositions, that is, propositions that are possibly true and possibly false.

This suggestion, however, misses the point of the current inquiry. The claim that it is not known a priori which propositions are true is either false or trivial. It is false if it includes conceptual truths, for it is known a priori that they are true; and it is trivial if it refers to contingent propositions. It is exactly this distinction, between conceptual truths and contingent propositions, that this inquiry attempts to understand. The truth-value of contingent propositions is indeed not known a priori. However, this is exactly because language leaves us with both options, and the question is: Why is this so? Claiming that contingent propositions are needed because their truth-value is not known a priori only begs the question.

Finally, it might be argued that the requirement that every conceptual possibility included in a language should be *actually* used in a true description of what exists is too strong. The need for contingent propositions for describing nature, for example, p, is explained by the fact that nature includes both the possibility of p and the possibility of not-p. Thus, according to this suggestion, the question I am asking receives a trivial answer. The need for language that includes unactualized conceptual possibilities is explained by the fact that nature includes unactualized conceptual possibilities. Another way to formulate this answer to the question I am asking in this chapter is in terms of "possible worlds." According to this interpretation, language must include conceptual possibilities because it ought to be able, not only to describe the actual world, but also all the possible worlds that are accessible from the actual world.

However, this answer to the question I am asking here is not only trivial but unfortunately also vacuous. It is vacuous because it is obvious that if nature includes unactualized conceptual possibilities, the language that is used for its description must include unactualized conceptual possibilities. Similarly, if the "possible world" terminology is adopted as an interpretation, or even an account, of modality, then the language that is used to describe the actual world ought to also be able to describe all the possible worlds. The question is, what is the difference between a nature that includes these possibilities, and a nature that does not? Alternatively, in the terminology of possible worlds, what constitutes the possible worlds which require the use of contingent propositions for their description? Hence, claiming that these possibilities are simply included in nature is evading the question, rather than giving an enlightening answer.

In passing, I note that the conclusion of my analysis provides an explanation for modality—specifically, for the idea that nature encompasses unactualized conceptual possibilities. Indeed, one might say that the investigation of contingency leads from modal contingency to what may be termed "temporal contingency," which refers to the possibility of assigning propositions conflicting truth-values depending on the temporal context of their use. This constitutes a reduction of modal contingency to temporal contingency (as discussed in Sect. 4.13), thereby accounting for the idea of "possible worlds."It can therefore be concluded, that if contingent propositions are required for describing what exists, this can only be explained if describing what exists requires the actual realization of the conceptual possibility of contingent proposition to be both true and false. That is, contingent propositions are required for describing something if and only if this description requires that these propositions will be both "true" and "false."

It is, however, doubtful whether the same proposition can be both true and false without a contradiction. I should emphasize in this context that this requirement does not imply that the same sentence, that is, the same structured signs in language, will express both true and false propositions. Rather, it is required that the same proposition shall be both true and false. The question of this possibility, and the necessary conditions for its actualization, is the task to which I turn my attention now.

In the following Sects. 4.3, 4.4, 4.5, 4.6, 4.7, 4.8, and 4.9), I develop the conceptual apparatus that enables a proposition to be both true and false without a contradiction. This apparatus will be eventually identified (in Sect. 4.10) with the temporal indexical expressions as what enables the same proposition to be both true and false, thus explaining the contingency of propositions in language. However, I emphasize that the following inquiry is undertaken without any reliance on our existing conceptual scheme.

Finally, I should clarify that my inquiry does not rely on the premise that temporal indexical expressions—or, for that matter, any other linguistic component, including the context operator I introduce as a technical tool—allow expressing the same proposition across different contexts, thereby enabling the ascription of conflicting truth values to the same proposition. This potential misunderstanding may arise because, given the difficulty of the subject matter and the complexity of the analysis, I rely on temporal indexical expressions throughout my inquiry to guide the reader through its direction and purpose.

Furthermore, not only is this a contentious semantic assumption, but it also entails a significant metaphysical commitment at the heart of the ongoing debate between A-theory and B-theory of time. In Chap. 7 I discuss these two theories of time in detail and explain why the conception of time developed here differs from both. Meanwhile, I wish to emphasize that the conclusions of my analysis should not be interpreted in light of A-theory. More specifically, despite a superficial similarity, my analysis does not assume any of the semantic, epistemological, or ontological foundations upon which A-theory is based.

To reiterate, the starting point of my analysis is the need for such a conceptual apparatus, and it is based on this requirement that I develop the underlying conceptual structure supporting this possibility. Moreover, even after associating this apparatus with temporal indexical expressions, I do not assume that these expressions inherently allow for this possibility. Instead, I demonstrate (in Sect. 4.10) that these expressions indeed enable the expression of the same proposition across different temporal contexts, thereby allowing for the assignment of conflicting truth values to the same proposition.

4.3 Conceptual Context

Nature requires the use of contingent propositions, that is, propositions that can be possibly true and possibly false, for its description. Contingent propositions are needed for description if and only if describing what exists requires the same proposition to be actually both true and false. My end is to expose the features that require, for their description, that the same proposition would be both true and false, thus exposing the necessary features of nature. However, the strategy I am employing to uncover these features is by asking: How is it possible for a proposition to be both true and false without a contradiction?

I should note that an answer to this question must be found in our language, as should be expected from the fact that it is used to describe what exists, and hence includes contingent propositions. I shall argue in Sect. 4.10 that, in our conceptual scheme, the explanation is found in indexical expressions, which enable the same proposition to have conflicting truth-values without a contradiction. More specifically, I shall argue that it is temporal indexical expressions—such as "past," "present" (or, "now"), and "future"—that enable the same propositions to be both true and false in different circumstances. However, I shall not rely on our actual conceptual scheme in the following investigation. Instead, I try to rebuild from nothing the conceptual apparatus that is needed for explaining the possibility of the same proposition being both true and false without a contradiction.

The reason for the strategy I employ is simple. Even if temporal indexical expressions enable the same proposition to be both true and false without a contradiction, it may be that this is merely one possible explanation for the use of contingent propositions for describing what exists, and hence does not represent necessary features of nature. Furthermore, our conceptual scheme is very complex and includes diverse conceptual connections. These connections, which obtain between the temporal indexical expressions and between them and numerous other distinctions, may or may not be themselves necessary features of nature. This conceptual complexity can obscure the nature of the specific conceptual mechanism that enables the same propositions to be both true and false without contradiction, and thus prevent a correct understanding of the necessary structure of nature.

As mentioned above, the difficulty is to understand how a proposition p, whose structure is yet to be determined, can be both true and false without a contradiction. Simply conjoining p and not-p results in a contradiction, as exemplified in proposition (1):

(1) $p \wedge \neg p$

Notice that p and $\neg p$ must be stated categorically. That is, it would not suffice to rely on, for example, a hypothetical proposition, that is, if q then not-p, as exemplified in proposition (2).

(2) $p \wedge (q \rightarrow \neg p)$

Proposition (2) does not imply both p and $\neg p$, thus successfully escapes a contradiction. However, although it uses a negation of proposition p, it does not actually deny proposition p. Hence, it cannot explain how the same proposition can be both true and false without a contradiction.

Although material implication does not solve the problem I am facing, it does suggest a path to a solution. What is in need is another conceptual operator, that I shall signify with the sign ©, whose exact meaning, that is, function, will be defined in what follows. This operator is introduced with the sole purpose of allowing us to categorically assert both p and $\neg p$ without a contradiction.

Simply adding another component to proposition p will not by itself solve the problem I am facing. That is, if the proposition that is both affirmed and denied is ©p, it is again impossible to avoid the contradiction, as demonstrated in proposition (3):

(3) $©p \wedge \neg(©p)$

To avoid contradiction, the new operator must not function as an ordinary distinction that is added to a proposition. To explicate its unique function, it is advisable to concentrate on its intended end, that is, to avoid a contradiction.

Suppose p is the proposition that is both affirmed and denied. To avoid the contradiction, it is necessary to affirm and deny this proposition in such a way that it would be impossible to infer both p and $\neg p$. This is possible only if the conceptual operator is used as a "containment" operator, that is, if it creates a conceptual space in which the contradictory propositions can be asserted, but remain isolated in such a way as would not allow extracting the contained propositions and infer that both p and $\neg p$.

I shall therefore try to use the conceptual operator that was conjectured earlier, as a means for containing p and $\neg p$, as demonstrated in proposition (4):

(4) $©p \wedge ©(\neg p)$

Notice the difference between propositions (3) and (4). While in proposition (3) the proposition which is both affirmed and denied is $©p$, in proposition (4) it is p that is both affirmed and denied. The conceptual component © functions as a containment operator, in the sense that it should not allow us to infer p and $\neg p$ from $©p$ and $©(\neg p)$ respectively. Neither should it allow us to infer $\neg ©p$ from $©(\neg p)$, which would again result in a contradiction. Thus, © can be used to allow the same proposition to be both affirmed and denied without a contradiction.

Although © can seemingly prevent a contradiction between p and $\neg p$, the question that is left unanswered is, how could proposition p and $\neg p$ both be true? For p is the same proposition in both its appearances in proposition (4), and hence it seems that if p is true, $\neg p$ should be false.

This difficulty should be explored at the most fundamental level of understanding of language. Let me begin with the basic idea, that a sentence is a combination of meaningful signs, which determines the proposition that is expressed by this sentence, that is, its description of what exists. There are several possible explanations for how the signs that comprise a sentence contribute to the determination of the truth-value of the proposition that is expressed by that sentence. These explanations are not mutually exclusive, and all should be considered:

(1) The usual convention in language is that each sign adds its independent contribution to the other signs it is conjoined with. Grammatical rules aside, it is possible to begin with one sign, for example, the predicate "red," which will be considered for the sake of the present analysis as a complete sentence, which expresses a true proposition if this possibility is realized, that is, if something red exists. To this sign, other signs can be added, each contributing another distinction, and thus each contributes to the determination of the truth-value of the proposition that is expressed. For example, adding "balloon" identifies the red object; this balloon can be ascribed to Jack; and so on. It makes no difference, according to this convention, with what constituent the constructing of this sentence begins. Each sign contributes to the already existing description, and the truth-value of the final proposition is determined by the combination of all these constituents. Each determination is hence independent of the other distinctions, which are added together to create the description. Each addition contributes to the final meaning of the sentence without ever changing the meaning of the signs it is conjoined with.

This suggestion cannot explain the possibility of *p* being both true and false when conjoined with ©. According to this suggestion, © is merely added to the meaning, that is, to the determination of the truth-value of *p*, which results in a new proposition ©*p*.

(2) A sign in a language can lack an independent meaning which relates to all its appearances in sentences. According to this suggestion, only in relation to the final combination of signs in a sentence can the meaning of an individual sign be determined. Although there may be something that is shared by all the instances of a sign, it is not strictly speaking a meaning. That is, its contribution to the meaning of different sentences can change, simply because it is conjoined with other components. Thus, adding the component © to "*p*" implies that the meaning of "*p*" changes.

This suggestion is problematic. It seems to introduce a superfluous complication to language, by postulating that the structure of sentences has little to no connection to its conceptual structure, including its connection with other sentences in a language. Nevertheless, it seems possible and perhaps even advisable to formulate such sentences in special cases. This is the case, for example, in a coded message, in which every occurrence of a sign can change its meaning, depending on an arbitrary set of rules that relates to its relations with other signs of the coded sentence.

Although this suggestion can explain how the sentence "*p*" can express propositions that have different truth-values when it is alone and then when it is conjoined with ©, it cannot explain contingency. This is because it implies that the content of the sentence "*p*" changes its meaning when conjoined with ©, and hence it expresses a different proposition. To explain contingency in language, the meaning of "*p*" must be preserved when conjoined with ©.

(3) According to option (1), the truth-value of ©p is determined "from the inside out," that is, *p* has a determined truth-value, and © merely adds its meaning to create a complex proposition, with a possibly different truth-value. The meaning of "*p*" is not affected by ©, and hence this cannot explain how proposition *p* can be both true and false. According to option (2), the meaning of "*p*" is determined by ©, and hence it can express true and false propositions, but only because they are different propositions. Hence, it seems that what is needed is a middle ground between these options. The meaning of "*p*" must not change, but the truth-value of *p* must be sensitive to the operator ©.

The truth-value of *p*, in the compound structure ©*p*, must therefore be determined "from the outside in." That is, it is impossible to determine the truth-value of *p* without first relating to the fact that it appears in the scope of ©. According to this suggestion, in the proposition ©*p*, © does not add an ordinary distinction to the proposition *p*. It instead enables determining a truth-value for *p*.

The conceptual operator © should therefore not only contain proposition *p*, but also create a context in which the truth-value of *p* is determined. In other words, *p* can receive different truth-values if it is sensitive to the containment operator. Notice, however, that it does not follow that the conceptual operator © that I have postulated is intended to fulfill two different functions. The functions of containment and of creating a context for determining the truth-value of a proposition are identical. I am merely refining the characterization of the conceptual component © and, in so doing, gaining an insight into its unique function. What now becomes clear is why proposition p is contained by the operator ©. It is because the determination of its truth-value depends on the context that is created by ©.

However, in light of this conclusion, the understanding of the component that is contained in the scope of the context operator © must change. For this use of © is possible if and only if the contained component is somehow sensitive to the context, that is, if there is a conceptual component in this proposition that is determined by ©.

It is important in this context to emphasize that a language should include the possibility of formulating propositions that are not sensitive to any context. More specifically, if it is impossible to make general claims about all contexts, then this view of language will be a victim of self-contradiction, to which all global relativistic views fall prey. For the claim that all propositions are context-dependent (that is, relative), which is supposed to be context-independent, would itself be context-dependent. Hence, it must be possible to make general statements, which are not sensitive to any context, either because they do not contain the component that is sensitive to the context, or because (if it is concluded that this component is somehow a necessary feature of all basic propositions in language) it is somehow neutralized by some conceptual mechanism.

I shall now try to apply the conclusions of this analysis to the difficulty at hand. My investigation began with p, which I assumed to be a complete proposition, that is, a component of language that is true or false independently of anything else. Recall that according to the use of language as a system of coordinates, there is no minimum number of coordinates that is needed to express a proposition. However, if more distinctions are added to a proposition, it changes. Hence, if p is a proposition, that is, if it has a determined truth-value, it seems that ©p expresses a different proposition. Thus, © fails to create a context in which the truth-value of p is determined.

It might be suggested that p expresses a complete proposition only because of a convention, according to which the implication of not including the conceptual context © is either its affirmation or its denial. Thus, although p is sensitive to a context, it is not necessary to state the conceptual component explicitly. However, my reason for not accepting this suggestion—that p implies a context without stating it explicitly—is that I wish to formulate a language in which the conceptual structure is reflected as clearly as possible in its sentential structure. Hence, if p includes any conceptual component that is sensitive to conceptual context, it must be reflected in its sentential structure.

Hence, since it was found that there is a need for propositions that are sensitive to context, this must be reflected explicitly in their structure. Thus, instead of p, the conceptual structure of the proposition in question is ©φ, in which φ stands for any combination of coordinates that are sensitive to context, and © for the component that determines the context. Conjoining © with the coordinates φ hence does not imply that another distinction is being added—the context operator is not an ordinary distinction that is used to describe what exists. The concept operator © merely determines a context for making these distinctions. Hence, φ must be interpreted as a combination of coordinates that do not yield a complete proposition by themselves. Instead, they require the use of a context operator for their determination.

In light of this analysis, it is necessary to update the structure of the propositions in question. Affirming and denying the same proposition now has the general structure that is demonstrated in proposition (5):

(5) $©\varphi \wedge \neg(©\varphi)$

However, again it is found that, in this simplified form, the contradiction is inevitable. I am again faced with the problem of how to avoid the contradiction, by finding a way to contain these propositions, in such a way that would not allow inferring both $©\varphi$ and $\neg(©\varphi)$. The way to solve this problem is, however, already paved, since the conceptual component © was introduced exactly for solving this problem, by creating a conceptual space that can contain propositions. It enables avoidance of the contradiction by separating the two contradictory propositions into different conceptual containments, as demonstrated in proposition (6):

(6) $©(©\varphi) \wedge (©\neg(©\varphi))$

However, this attempt to escape the contradiction will not do. For now it is understood that this containment is achieved by creating a context in which distinctions are made. In light of this understanding, it is impossible to ignore the fact that the two conjuncts in proposition (6) share the same context operator ©, and hence it seems that proposition (6) implies proposition (7), which is inconsistent:

(7) $(©)(©\varphi \wedge \neg(©\varphi))$

To avoid this inconsistency, different contexts must be used. Fortunately, there is no need to introduce another independent context operator. It is possible to use the original conceptual context © and its negation, that is ¬©, for, as was shown in Chap. 2, every conceptual component in a language must be capable of being negated. There are therefore two possibilities for solving the problem, as demonstrated in propositions (8) and (9)

(8) $©(©\varphi) \wedge ((\neg©)\neg(©\varphi))$
(9) $(\neg©)(©\varphi) \wedge (©\neg(©\varphi))$

Both options are acceptable for solving the problem of inconsistency. Furthermore, there is no need to decide between them because they are both implied by the introduction of the conceptual operator. In other words, there is nothing to choose between them. For simplicity, I shall refer only to proposition (8) in the following analysis.

A few explanatory remarks regarding the meaning of proposition (8) are in order. To begin with, I should emphasize that the inner parentheses in each of the conjuncts of proposition (8) are important for structuring these propositions. If it was not for this structure, the conjuncts of proposition (8) would be interpreted as a straightforward combination of coordinates, that is, they would be understood as "©©φ" and "(¬©)¬©φ." The problem is that in the first proposition in this conjunction, that is, "©(©φ)," the same distinction is being repeated, and hence one of the appearances of "©" seems to add nothing to the meaning of the proposition. The second proposition in this conjunction, that is, "(¬©)¬©φ," includes a negation sign that is attached to the second appearance of "©." If the second negation sign refers to the following context operator ©, its structure is "(¬©)(¬©)φ," in which again the same context operator is repeated again. In this case, the former and the latter conjuncts in proposition (8) refer to different propositions. Hence, this interpretation is unable to explain the possibility of both affirming and denying the same proposition.

Another important thing to notice is that the spatial order of the conceptual operators © in this proposition is of conceptual significance, that is, it is a conceptual component of these sentences. In fact, the parentheses function as a means for securing the unique position of each of the instances of ©. Recall in this context that the order of the component of a sentence is not always conceptually important (although it may be grammatically important). Thus, there seems to be no meaning to the order of the words in the sentence "Moses existed," and the only reason why this order is preferred to "existed Moses" is grammatical. In contrast, the use of spatial distinctions in sentence (8) is important. It enables repetition of the conceptual operator in a structured way. Furthermore, this allows us to give meaning even to a reapplying of the same distinctions, that is, (©….©(©(©φ))) or ((¬©)….(¬©)((¬©)((¬©)φ))). This possibility will be found important in what follows.

Returning to the investigation, notice that the first instance of the conceptual operator © functions as a sentential connective in each of the conjuncts of proposition (8). That is, it is conjoined with a proposition, that is, ©φ. The fact that ©φ is a complete proposition implies that it has a determinate truth-value. Nevertheless, it can be inserted into the context of another context operator.

This use of the conceptual operator © as a sentential connective is expected given the previous considerations. That is, it was expected to create a context for itself, thus creating a second-order context. To begin with, if a context operator can be coupled with a combination of coordinates in language, then there is no reason why the context operator could not be coupled with a combination of coordinates that constitute a proposition, as long as this proposition is sensitive to a context. Furthermore, the conceptual operator © was introduced to explain the possibility of the same proposition being both true and false, and this is only possible if it can create a context for complete propositions.

I conclude that the conceptual operator © can allow the same proposition to be both true and false without a contradiction. As propositions (8) demonstrates, it is possible to formulate a description in which the same proposition, that is, ©φ, is both true and false, without this description implying a contradiction. This is because the truth-value of ©φ is sensitive to a context, and hence it cannot be extracted from this context without affecting its truth-value.

The discussion so far has been purely formalistic. However, I now wish to explore the different ways in which this conceptual formulation can be interpreted. For this purpose, it would be not only permittable, but also advisable, to rely on our language, and inquire, based on the conceptual apparatus we possess, how such a conceptual structure can be explained.

Generally, when a component is added and refers to a complete sentence, several possibilities should be clearly distinguished:

(1) Although the addition relates to a complete sentence, it merely adds another distinction to the description made by the original sentence. Examples of such cases in our language are "I know that *p*" and "It is necessary that *p*," or simply "$q \wedge p$."

In this case, however, the original proposition follows from the compound proposition. Hence, this option cannot explain how the same proposition can be both affirmed and denied without a contradiction.

(2) The addition can refer to the content of the original sentence. In our conceptual scheme, this is demonstrated, for example, by psychological propositional attitudes, such as "I believe that *p*."

Although it is impossible to infer *p* from these compound propositions, this is only because the truth-value of the original proposition is irrelevant to the determination of the truth-value of the compound proposition. Hence, these cases are unable to explain the possibility of ascribing different truth-values to the same proposition.

(3) The closest example in our language of a conceptual component of the type hypothesized here is found in indexical expressions, such as "here," now," and so on. Such expressions are usually interpreted as components of sentences. This is obviously true since they can replace temporal distinctions. Thus, for example, instead of saying "I shall meet you on the 7th of July," one can simply say, "I shall meet you now." However, there is also a use of indexical expressions to create a context for complete sentences, which are sensitive to this context, as exemplified in the sentence "The claim that 'the store is open now' is now no longer true." This function of temporal indexical expressions fits into the characterization of the conceptual operator ©. However, I shall not rely on this similarity in the following analysis, in order not to presume that it is the correct or only possible solution, to the difficulty at hand.

4.4 High-Order Uses of the Context Operator

The previous section concludes that affirming and denying the same proposition without a contradiction is possible using a conceptual operator ©, which both gives context for the determination of the truth-value of conceptual coordinates and contains the resulting proposition. What is required now is to give a more precise meaning to this conceptual operator, that is, to define its exact function in language. My aim in this section is to offer a preliminary refinement of the use of this operator, by determining more accurately the conventions that govern its use in high-order contexts, that is, when it is applied to itself.

To begin with, it should be noted that the idea of a conceptual context, and with it the idea of sensitivity to a context, raises the danger of infinite regress. For I have shown that to explain how a proposition can be both affirmed and denied without a contradiction, the conceptual context operator © needs to be applied to itself, that is, as a second-order context. It should be remembered, however, that the idea of a context operator relied on the assumption that distinctions in a language require context. Otherwise, it would have been impossible to explain why the truth-value of ©φ is different from the truth-value of (¬©)φ. Hence, if © is also sensitive to context, and it is possible to reapply this operator to itself, that is, ©(©φ), then it seems that not only is it possible, but indeed necessary, to reapply this operator again and again indefinitely. For if φ is sensitive to a context, and hence requires a context to determine a truth-value, so does ©φ, which equally requires a context for determining its truth-value, and so on ad infinitum. This results in a vicious regress because at each stage another context is required to formulate a complete sentence, that is, to express a proposition with a definite truth-value.

To avoid an infinite regress, it is hence necessary to refine the conceptual context operator ©, by introducing a convention that governs the interpretation of sentences such as "©φ". Prima facie, two options are available. The first option is to determine a default context operator. That is, to introduce the convention that the conceptual context of sentences of this type is either © or (¬©). This could be decided arbitrarily or according to the most inclusive context operator that appears in a sentence. Thus, according to this interpretation, if © is chosen as the default interpretation, ©φ is logically equivalent to ©(©φ), as well as to ©(©(©φ)), and so on. The second option is to interpret any sentence that includes a conceptual context operator as a general sentence, that relates to a disjunction of each of the possibilities, that is, ©¬(¬©). According to this suggestion, ©φ is logically equivalent to (©¬(¬©))(©φ).

The latter option seems appealing, in light of the usual convention in natural language, that the nonappearance of a distinction implies that it is neither affirmed nor denied. However, it is easy to show that this suggestion is not viable. To see why, take any proposition that includes a context operator, for example (10):

(10) ©φ

According to this suggestion, proposition (10) is logically equivalent to proposition (11):

(11) (©¬(¬©))(©φ)

However, to determine its truth-value, it is necessary to rely on the fact that proposition (11) implies proposition (12), and examine the truth-value of each of its disjuncts:

(12) ©(©φ) ∧ (¬©)(©φ)

However, to determine the truth-value of each of the disjuncts in proposition (12), it is again necessary to apply the suggested convention. For, according to this suggestion, the first disjunct in proposition (12) should be interpreted as demonstrated in proposition (13):

(13) ((©¬(¬©))(©(©φ))

Again, proposition (13) depends on the truth-value of a disjunction, each of whose disjuncts needs further interpretation, and so on ad infinitum.

After eliminating the second option, we are left with the first option, which relies on the idea of a default context operator. I begin the discussion with propositions whose context operator is ©, and only at a later stage turn my attention to propositions whose context operator is (¬©).

To begin with, it is possible to quickly reject one rule for determining the default context operator. According to this rule, the context is always changing. That is, ©φ is logically equivalent to (¬©)(©φ), as well as to ©((¬©)(©φ)), and so on. This suggestion again involves an infinite regress, which would not allow ascribing a truth-value to any such proposition, because at each higher level the context operator changes. The default context operator must therefore be constant, that is, either © or (¬©). However, the choice between them is arbitrary since I have yet to define © and (¬©) in such a way that allows making a distinction between them. Hence in what follows I choose to continue with ©.

It can be concluded that there is only one option to determine the default context operator of ©. According to this convention, propositions of the type "©φ" are logically equivalent to ©(©φ), ©(©(©φ)), and so on. This convention allows ascribing a truth-value to the basic proposition ©φ without descending into an infinite regress. That is, this repetition of the context operator must be vacuous.

It can therefore be concluded that proposition (8), which explains the ability to both affirm and deny the same proposition without a contradiction, is logically equivalent to proposition (14).

(14) (©φ) ∧ ((¬©)¬ (©φ))

So far, I have concentrated on the use of the context operator as a sentential connective, that is, in cases in which it applies to whole sentences. For these are the uses that are most relevant for the present inquiry, whose main aim is to understand how this operator can enable affirming and denying the same proposition without a contradiction.

However, in its use to create a context for itself, the context operator must also have another use, due to the fact that it is already established that it can both be an integral part of a proposition and relate to a complete proposition. The two uses of the context operator are demonstrated in the sentence "(¬©)(©φ)." I have shown that this implies that the conceptual context operator © itself is sensitive to a context. Otherwise, it would have been impossible to explain how the context component (¬©) can have any effect on the determination of the truth-value of ©φ—a fact that was found to be necessary for explaining how the same proposition could be both true and false. For if the coordinates of φ are determined according to ©, while © itself is not sensitive to any context, then adding (¬©) as a sentential connection is irrelevant for determining the truth-value of the inner proposition.

Hence, it is possible to distinguish between two second-order uses of the conceptual context operator. The first use is its already identified use, in which it is applied to a complete proposition which includes itself, as exemplified in proposition (15):

(15) $(\neg©)(©\varphi)$

The second use is when it is used to create a context for another context operator, as exemplified in proposition (16):

(16) $(\neg©(©))\varphi$

The difference between the two propositions is that, while in proposition (15) the context operator (¬©) determines a context that relates to proposition ©φ as a whole, in proposition (16) the context operator (¬©) determines a context for the context operator ©, and the resulting operator determines a context for the coordinates that are included in φ.

Notice that I am not claiming that propositions (15) and (16) are not logically equivalent. This is yet to be determined when the semantic properties of the context operator will be further explored and defined. What I am arguing is that the two sentences do not share the same structure, and this possible difference should be further studied.

Because © is the default operator, any proposition of the type ©φ is logically equivalent to ©(©φ). Hence, proposition (15) is logically equivalent to proposition (17).

(17) $(\neg©)(©(©\varphi))$

To understand why, recall that in proposition (17) the truth-value is determined "from the outside in." That is, (¬©) determines the context of the second-order operator ©, and the result determines the context for the proposition ©φ. Hence, when the context operator (¬©) creates a context for the default second-order operator ©, the latter operator becomes "transparent," and the resulting context operator is simply (¬©).

This analysis of the higher-order use of the context operator further implies that proposition (16) is logically equivalent to proposition (18):

(18) $(\neg©)\varphi$

The reason is that the context operator © was identified as the default context. Now, while not all the higher-order uses of the conceptual operator have been determined, in proposition (16) the (¬©) operator creates a context for the default context operator, that is, ©. As the analysis of propositions (15) and (17) shows, in this case, that the resulting context operator is (¬©).

Before concluding this section, I should note that this line of thought seems to suggest that the context operator © can always be eliminated when it is in the direct scope of the context operator (¬©), and this also applies to proposition (15), that is, (¬©)(©φ). which according to this conclusion is logically equivalent to proposition (18), that is, (¬©)φ. However, I shall not pursue this suggestion further. This is because, as shall be evident in what follows, it has significant implications for the ability to ascribe different truth-values to the same proposition without a contradiction.

4.5 The Conceptual Structure of the Context Operator

In this section, I wish to further define the properties of the context operator ©. While exploring this subject, it must be kept in mind that the context operator © is intended to allow both affirming and denying the same proposition. That is, it should make it possible to express both proposition (a) and proposition (b) without a contradiction.

(a) $©\varphi$
(b) $(\neg©)\neg(©\varphi)$

To avoid a contradiction, proposition (b) must not imply proposition (c).

(c)$\neg(©\varphi)$

The question I am asking in this investigation is: what are the rules that govern the use of the context operator © that coherently allow this? As will be evident in what follows, there are constraints on the use of the context operator, especially regarding its higher-order uses.

Before beginning the inquiry, I must address the abstract nature of this discussion and the difficulty of understanding its significance. Some might even wonder whether the questions I raise here are meaningless, given the fact that this discussion is purely syntactical, and no meaning has been given to the context operator. However, I should emphasize that this question is of the utmost importance for uncovering the necessary features of nature. As will be evident from the conclusions of this chapter, the distinction between © and (¬©) is a temporal distinction. Hence, the conceptual features of these distinctions describe the structure of time. The question of whether there is only one default context operator is thus relevant to the resulting structure of time, and specifically to the question of whether describing what exists from a point of view in time, which in our conceptual scheme is the present, is a necessary feature of nature.

I begin with the simplest propositions that can be formulated by the context operator ©, that is, $©\varphi$ and $(\neg©\varphi)$. These propositions must be contingent, as required by the conditions for the meaningfulness of expressions in language, which were revealed in Chap. 2. Apart from these basic propositions, I examine every sentence that can be formulated by a second-order use of the context operator and its negation, which is the level of complexity that is required for formulating proposition (b).

Proposition (b) includes two basic conceptual components: "φ" and "©." According to the conception of language as a system of coordinates, each component can be negated, which adds "¬φ" and "¬©." However, I ignore "¬φ" in what follows because it is irrelevant to the present analysis. Furthermore, propositions can also be negated, which adds more possibilities for formulating sentences using these constituents.

There are 12 possible formulations at the level of complexity that is reflected in proposition (b), and these possibilities are presented in Table 4.1.

Notice that proposition (4.1.1) is identical to proposition (a); proposition (4.1.11) is identical to proposition (b); and proposition (4.1.3) is identical to proposition (c). Hence, the context operator should be defined in such a way that allows proposition (4.1.1) and proposition (4.1.11) to both be true; and to allow that without a contradiction, proposition (4.1.11) must not imply proposition (4.1.3).

To begin with, it should be noted that the conclusion of Sect. 4.4, which identifies © as the default context operator, implies that propositions (4.1.1) and (4.1.5) are identical, as well as propositions (4.1.3) and (4.1.7).

Following the same line of thought, one may wonder if propositions (4.1.2) and (4.1.6), as well as propositions (4.1.4) and (4.1.8), should not be identified as identical. This is because the only difference between the two propositions in each pair is the addition of the default context operator ©. Recall, however, that I have identified the default context operator only for propositions in which the context operator is ©, that is, propositions of the general structure ©φ. This makes sense, because in these cases the default operator is identical to the actual operator of the proposition.

It might be suggested that similar considerations apply to propositions in which the context is (¬©), that is, propositions of the general structure (¬©)φ, in which the default context operator should likewise be (¬©). According to this suggestion, propositions (4.1.2) and (4.1.10) should be identified as identical, as well as propositions (4.1.4) and (4.1.12).

Let me therefore explore the suggestions that the default context operator, in propositions of the general structure (¬©)φ, should be (¬©). An important implication of this suggestion is that the actual conceptual apparatus in propositions that include context operators is limited to the basic context operator and the number of orders of contexts above it, because in each higher-order context there is only one meaningful option. That is, if the general structure of a proposition is (¬©)φ, the

Table 4.1 The complexity that is reflected in proposition (b)

4.1.1			©	φ
4.1.2			(¬©)	φ
4.1.3		¬	©	φ
4.1.4		¬	(¬©)	φ
4.1.5	©		©	φ
4.1.6	©		(¬©)	φ
4.1.7	©	¬	©	φ
4.1.8	©	¬	(¬©)	φ
4.1.9	(¬©)		©	φ
4.1.10	(¬©)		(¬©)	φ
4.1.11	(¬©)	¬	©	φ
4.1.12	(¬©)	¬	(¬©)	φ

only meaningful addition of a context operator to this proposition is ©; while if the general structure of a proposition is ©φ, the only meaningful addition of a context operator to this proposition is (¬©).

To understand the implications of this suggestion, I begin with each of the two basic possibilities, that is, ©φ and (¬©)φ. These options are the first-order use of the context operator, as presented in the first row in Table 4.2.

Interestingly, the second-order use of the context operator does not increase the number of possible formulations, which again amounts to two, as described in the second row of Table 4.2. This is also true for every higher-order use of the context operator. Although at each stage the context operator © or its negation can be added to the lower-order proposition, because one of them is the default context operator, only its negation can be meaningfully added to the lower-order proposition. That is, proposition ©(©φ) is logically equivalent to ©φ, and hence the context operator © does not contribute to the meaning of the sentence "©(©φ)." On the other hand, the proposition (¬©)((¬©)φ) is logically equivalent to ((¬©)φ), and hence the context operator (¬©) does not contribute to the meaning of the sentence "(¬©)((¬©)φ)."

Thus, in each higher-order use of the context operators, there are only two options, therefore this conceptual structure results in just two columns. The same conceptual structure that is reflected in Table 4.2 can be more simply and accurately depicted by stating the basic context operator and the number of orders, as exemplified in Table 4.3.

The resulting conceptual structure therefore amounts to two different dimensions, one designated by the context operator © and the other by its negation. Further, notice that there is no connection between the two dimensions, nor is there any logical connection between the different cells in each column. This is because the context operator is not a truth-functional connective. That is, the truth-value of ©*p* is not a function of the truth-value of *p*, because the truth-value of *p*, while in the scope of ©, is partly determined by ©.

Hence, it might be argued that the resulting conceptual structure is not one in which a context determines a truth-value. The truth-value of a proposition is determined by an ordinary combination of signs. This is comparable to two-dimensional space, in which the context operator and its negation are used as a coordinate on one of its dimensions, which includes only two possibilities, and the other dimension can be mapped with the help of integers. Suppose that this space includes two different colors, black and white, then mapping this space requires only three coordinates, for example, $©^1$B, $(¬©)^1$W, $©^2$B, $(¬©)^2$W, and so on. There is no need to ascribe different truth-values to any proposition to describe this space.

Table 4.2 High-order use of the context operator – first suggestion

First order	©φ	(¬©)φ
Second order	(¬©)(©φ)	©((¬©)φ)
Third order	©((¬©)(©φ))	(¬©)(©((¬©)φ))
…	…	…

Table 4.3 High-order use of the context operator – first suggestion simplified

First order	©φ	(¬©)φ
Second order	$©^2$φ	$(¬©)^2$φ
Third order	$©^3$φ	$(¬©)^3$φ
…	…	…

Recall that my inquiry is directed by an attempt to explain how proposition (a), that is, ©φ, and proposition (b), that is, (¬©)¬(©φ), can be both true without a contradiction. According to this suggestion, proposition (b) should be translated as ¬($©^2$φ). This is because, according to this interpretation, there is no connection between the two propositions; each is understood as relating to another fact. Furthermore, proposition (b) does not imply proposition (c), that is, ¬(©φ), as stipulated. This is because of the trivial fact that it seems to relate to another fact altogether.

This interpretation of the context operator raises difficulties. However, it does not imply that I have reached a dead end. There may nonetheless be a way to explain the need for ascribing different truth-values to the same proposition according to this interpretation of the context operator. However, the fact that this suggestion leads to a two-dimensional conceptual space, in which one dimension plays no role in explaining the need to ascribe different truth-values to the same proposition, suggests that this route fails to expose the *necessary* features of nature. This conceptual apparatus seems redundant for explaining the contingency of propositions in language.

As will be clear in what follows, the suggestion that the default context operator is always the same leads to a simpler, one-dimensional interpretation. Hence, in my quest for the *necessary* features of nature, the one-dimensional interpretation of the required conceptual structure should be preferred.

There is, however, another concern regarding the suggestion of two default context operators. This is the worry that this suggestion does not describe a single unified nature. To understand this concern, recall that the context operator determines how the distinctions that describe nature ought to be applied, and that it is impossible to extract the propositions from the scope of the context operators. Furthermore, the propositions that are sensitive to context are supposed to be the basic propositions that describe what exists, upon which all other meaningful propositions rely. However, if at the most basic level two distinct contexts are found, which have no logical connection between them, the suggestion that this conceptual structure can describe a single unified nature seems unjustified.

A natural response to this concern would be to resort to logic, by relying on the fact that propositions (a) and (b) imply their conjunction, that is, proposition (14).

(14) (©φ) ∧ ((¬©)¬(©φ))

However, the fact that proposition (14) is an implication of the two separate conjuncts fails to indicate a unification. Furthermore, it can be used to show that there is no context from which this conjunction can be described, because their default context operator is different. Hence, proposition (14) is not logically equivalent to either of propositions (19) and (20), which are the only two options for unifying the two conjuncts in the scope of a single context.

(19) ©((©φ) ∧ ((¬©)¬(©φ)))
(20) (¬©)((©φ) ∧ ((¬©)¬(©φ)))

Hence, the different context operators may just as well refer to different realities, which in Table 4.3 are represented by the two separate columns.

In summary, if the basic propositions in a language require a context for their determination, their conjunction, as a description of the same nature, must share the same context. Hence, a shared context operator to unify the conjuncts in proposition (14) is needed. That is, there must be the same default context operator for unifying all the context-dependent propositions in a language. According to my previous stipulation, this is the default context operator ©. Proposition (14) must therefore be logically equivalent to proposition (19), in which both (a) and (b) are located in the scope of the same context operator ©.

It can therefore be concluded that the default context operator, which is shared by all context-sensitive propositions, is ©. Hence, propositions (4.1.2) and (4.1.6) in Table 4.1 are logically equivalent, as well as propositions (4.1.4) and (4.1.8). Furthermore, because the context operator © is stipulated to be the default context operator, its addition to a sentence cannot contribute to the determination of the truth-values of the propositions that are expressed by the resulting sentence. Hence, in order not to render the context operator meaningless, adding its negation must contribute to the meaning of the sentences. This implies that propositions (4.1.1)—(4.1.4) are not logically equivalent to propositions (4.1.9)—(4.1.12) respectively.

4.6 Negation and the Context Operator

It is now time to examine proposition (b) more closely.

(b) (¬©)¬(©φ)

In proposition (b), proposition (a) "(©φ)" is located in the scope of the context operator (¬©), resulting in a compound proposition. There are, in natural language, compound propositions in which a basic proposition *p* is conjoined with other expressions, for example, "I know that *p*," or "I believe that *p*." There is, however, a fundamental difference between the latter examples and proposition (b). In the latter examples, the truth-value of *p* is not affected by the expression with which it is conjoined. In proposition (b), on the other hand, the truth-value of (©φ) is supposed to be sensitive to the context operator, and hence it can be different from its truth-value when it is not in the scope of (¬©).

The question is, how can one give meaning to the claim that the truth-value of ©φ can be different when it is in the scope of (¬©)? The context operator (¬©) was introduced to contain ©φ, which means that it is impossible to extract ©φ from the scope of (¬©) without affecting the way its truth-value is determined. How can its truth-value be identified while it is in the scope of (¬©)?

An example can help to understand this problem. Suppose you are told that the compound proposition "I know that ¬*p*" is true. You can infer the truth-value of *p* because the truth-value of the compound proposition depends on the truth-value of *p*, which in turn is independent of its appearance in the scope of the propositional attitude "I know that…" On the other hand, it is impossible to infer, from the truth-value of the compound proposition "I believe that ¬*p*" anything about the truth-value of *p*. However, its truth-value can be determined by examining *p* outside of the context of the propositional attitude "I believe that…".

The problem with claiming anything about the truth-value of (©φ) in the scope of (¬©) is that if one speaks of the compound proposition, one seems to be unable to speak about the truth-value of ©φ as an individual proposition; while if one speaks of the truth-value of ©φ as an individual proposition, it is no longer in the scope of (¬©). How can one determine the truth-value of ©φ as an individual proposition while it is in the scope of (¬©)?

It might be suggested that the fact that ©φ is supposed to be false in the scope of (¬©) is reflected in the fact that it is negated. This is true. However, this negation is also in the scope of the context operator (¬©). Rather than solving the problem, I am putting forward here, this leads to the same problem, since it is impossible to extract the proposition ¬(©φ) from the scope of (¬©).

The only way to be able to determine the truth-value of ©φ, while it is in the scope of (¬©), is if the truth-value of the compound proposition directly reflects the truth-value of (©φ), that is, (¬©)(©φ) is true if and only if ©φ is true in its scope of (¬©). Hence, the claim that ©φ is true in the scope of (¬©) amounts to the claim that (¬©)(©φ) is true; while the claim that ©φ is false in the scope of (¬©) amounts to the claim that (¬©)(©φ) is false.

I should note that this is not only a requirement that arises from the need to be able to characterize the truth-values of propositions while they are in the scope of the context operator. It is also expected from the way I have defined the context operator, as a conceptual component that does not add a distinction to the propositions which are in its scope. This characterization thus also implies that the truth-value of the compound proposition, that results from applying the context operator to a basic proposition, can only be determined by the truth-value of the basic proposition while it is in the scope of the context operator.

This analysis concludes that the context operator is "transparent" to truth-values, and hence to also to negations. This implies that the truth-value of the most basic proposition in a compound proposition that includes several context operators can be identified by the truth-value of the compound proposition and by adding all the negations that are included in this proposition. Thus, according to this analysis, proposition (b), that is, (¬©)¬(©φ), is logically equivalent to ¬(¬©)(©φ), and implies that the truth-value of ©φ, in its scope, is "false."

Table 4.4 The conceptual structure that supports proposition (b)

4.4.1		©	φ
4.4.2		(¬©)	φ
4.4.3	©	©	φ
4.4.4	©	(¬©)	φ
4.4.5	(¬©)	©	φ
4.4.6	(¬©)	(¬©)	φ

Although I have yet to identify the context operator with anything in natural language, I should point out that my conclusion also fits with the temporal indexical expressions. For example, the claim that "in the past, it was false that 'it is raining now'" is logically equivalent to the claim that "it is false that 'in the past, it was (true that) 'it is raining now''." Temporal indexical expressions are therefore "transparent" to negations, similar to the context operators.

An implication of the conclusion that the context operator is transparent to negations is that Table 4.1 need not include the negation column, because it does not represent an additional conceptual apparatus. Hence, Table 4.1, which is intended to demonstrate the conceptual structure that supports proposition (b), can be reduced to Table 4.4.

For clarification, recall that propositions (4.4.1) and (4.4.3) are logically equivalent, as well as propositions (4.4.2) and (4.4.4).

Furthermore, it should be noted that negation of a context operator, that is, (¬©), serves a different function than that of negations of complete propositions, and hence the previous rule of transformation does not apply to them. That is, the inequivalences that are described in propositions (21) and (22) hold.

(21) ¬(©φ) ≡/≡ (¬©)φ
(22) ¬((¬©)φ) ≡/≡ ©φ

This is important because the inequivalence between propositions (a) and (b) must hold, as demonstrated in proposition (23).

(23) ©φ ≡/≡ (¬©)¬(©φ)

For if proposition (a) is logically equivalent to proposition (b), it implies a failure to ascribe different truth-values to proposition ©φ.

The explanation for the difference in the uses of negation is simple. While the discussion in this section focused on the use of negations as sentential connectives, in (¬©) the negation is used to deny the context operator per se, and hence this expression denotes a context different from ©. This use of negation marks a necessary condition for the meaningfulness of the context operator, and hence reflects a necessary conceptual apparatus. It signifies a different possibility on the fundamental conceptual dimension on which the context operator © is located.

4.7 The Structure of the Context Operator Series

My investigation is directed by an attempt to avoid a contradiction between propositions (a) and (b).

(a) ©φ
(b) (¬©)¬(©φ)

The conclusion of the analysis of the previous section, regarding the use of negation in the scope of a context operator, was that proposition (b) is logically equivalent to proposition (b').

(b') ¬((¬©)(©φ))

Furthermore, the identification of the context operator © as the default context operator implies that adding it to a complete sentence cannot contribute to the determination of the truth-value of the proposition that is expressed by the resulting sentence. Adding it to a sentence is only a way to explicitly state what is conceptually implicit.

The fact that the context operator © is the default context operator, and hence can always be added without affecting the truth-value of the proposition that is expressed, implies that it can always be similarly omitted. Thus, this conclusion implies that proposition (b') is logically equivalent to proposition (b"):

(b") ¬((¬©)φ)

This further implies that proposition (4.4.5), in Table 4.4, is equivalent to propositions (4.4.2) and (4.4.4). This reduces the conceptual structure that is enabled by the context operators to the series described in Table 4.5, which describes the conceptual possibilities that are included in each higher-order application of the context operator.

The resulting conceptual structure hence amounts to a linear series, in which the higher-order uses of the context operator are defined with the help of the negation of the default context operator. There are conceptual connections between the propositions in this series, as each higher-order proposition is constructed based on the conceptual structure of the proposition of the lower order. Notice that this is also true regarding the second order. This is because, as the previous analysis shows, (¬©)φ is equivalent to (¬©)(©φ). However, notice that proposition (4.5.1) remains unique in this series of propositions, because its context operator is different from the other propositions in this series.

Table 4.5 Higher-order application of the context operator

4.5.1	©φ
4.5.2	(¬©)φ
4.5.3	(¬©)((¬©)φ)
4.5.4	(¬©)((¬©)((¬©)φ))
	...

It is natural to interpret the higher order proposition, such as proposition (4.5.3), that is, (¬©)((¬©)φ)), as the result of inserting a lower-order proposition (4.5.2), that is, (¬©)φ) to the scope of a new context operator (¬©). However, there is nothing that can rule out a different interpretation, according to which it is the result of adding a new context operator (¬©) in the scope of the context operator (¬©) of proposition (4.5.2). That is, there is nothing in the resulting proposition that can indicate whether the newly-added context operator (¬©) should be identified with the left or the right context operator in proposition (4.5.3).

Indeed, according to the last suggestion, it is possible to interpret all the propositions from (4.5.2) and higher as variations of a basic proposition (24), in which occurrences of the default context operator © are replaced with its negation, that is, (¬©).

(24) (¬©)(©©©©©…φ)

4.8 Logical Considerations in High-Order Use of the Context Operator

There remains, however, the question of the logical connections between the members of this series. So far, the only limitation that was found in the stipulation that proposition (b), which is logically equivalent to the negation of proposition (4.5.2), should not contradict proposition (a), which is equivalent to proposition (4.5.1), that is, they should be consistent. This implies that proposition (4.5.1) does not entail proposition (4.5.2).

There remains, however, the question of the logical connections that do obtain between the propositions of Table 4.5. A possible interpretation of this conceptual apparatus would be to identify the conjoined context operators (¬©) as a simple way to enumerate the members of this series, thus determining their location in this series. According to this interpretation, the conceptual connections that obtain between complex context operators amount to simple relations of linear order, which hence can be replaced by numbers, that is, (¬©)φ ≡ 2φ, (¬©)((¬©)φ) ≡ 3φ, and so on.

The interpretation of the logical relations between the members of this series has important implications. Recall that this series is supposed to reflect a fundamental feature of nature. It could imply, for example, a discrete interpretation of this dimension (which will be later identified with time). It could also imply logical atomism, because there are no logical connections between the different members of this series. Hence, it is of extreme importance to ascertain the logical connections between the members of this series.

However, the interpretation suggested above is false. There must be some type of logical connection between the propositions in this series. This is because if there are no logical connections between the different propositions in this series, then the conceptual structure that is described in Table 4.5 is superfluous, and as demonstrated in Table 4.3, there is no need to ascribe different truth-values to any proposition to describe this conceptual space. On the other hand, a straightforward logical equivalence would render the whole conceptual structure that is described in Table 4.5 meaningless.

A clue for revealing the connections between the propositions in Table 4.5 is the fact that each higher-order proposition relies on the conceptual structure of the lower-order proposition, that is, a proposition that includes fewer context operators. This hints at the existence of logical connections between propositions of different orders. Furthermore, this observation is supported by the conclusion that the context operators are "transparent" to negations. Taken together, these facts suggest that the truth of the higher-order propositions somehow relies on the truth of the lower-order propositions.

The only seemingly straightforward logical connection that can be established between the different members in this series is that a proposition is implied by either a higher-order proposition or a lower-order proposition. However, the truth-value of a lower-order proposition determines the truth-value of a higher-order proposition. This is because it conflicts with the idea of a context operator, which is supposed to affect the truth-value of the propositions in its scope, whose distinctions are supposed to determine the truth-value of the compound proposition. Furthermore, because a higher-order proposition is created by adding a conceptual operator to a lower-order proposition, that is, it includes another conceptual coordinate, it is expected that a lower-order proposition will be interpreted as more general. Hence, according to this interpretation, the truth of a proposition of a higher order implies the truth of a proposition of a lower order.

Again, notice that this does not apply to the relation between proposition (4.5.1), that is, ©φ, and proposition (4.5.2), that is, (¬©)φ. This is because of the change in the context operator, which, due to the uniqueness of the context operator, can change the truth-value of the proposition that is within its scope. All the propositions except (4.5.1), on the other hand, share the same context operator (¬©), and hence it is expected that adding the same context operator would function just like adding another distinction to a proposition, that is, making it more specific. Hence, just as the proposition "Jack has a red balloon" implies the more general proposition "Jack has a balloon," it seems reasonable to expect that the lower-order propositions follow from the higher-order propositions in Table 4.5 (again, excluding proposition 4.5.1, which has a different context operator).

This brings me to the meaning of the context operator (¬©). If proposition (4.5.2), that is, (¬©)φ, follows from propositions of higher order in this series, it should be understood as a general term, in the sense that it ranges over all the propositions that are determined by the context operator (¬©). Hence, it can be interpreted as generally claiming that, under the context operator (¬©), φ is realized.

This brings me to another important point that must be kept in mind. I began my investigation by postulating a context operator ©. The conditions for the meaningfulness of expressions in language, as revealed in Chap. 2, imply that there must also be a complementary context operator (¬©). Furthermore, in the course of this analysis I have found an increasing need to rely on the context operator (¬©) to explain the possibility of ascribing conflicting truth-values to the same proposition without a contradiction. Recall, however, that negating a coordinate does not necessarily denote a specific possibility on a conceptual dimension. On the conceptual dimension of colors, for example, not-red is a general coordinate, which denotes all the colors that are not red. Similarly, the conceptual dimension in which © is located is yet to be determined, and it may contain more than two possibilities. That is, there may be more context operators than the two I have identified so far.

4.9 Second-Order Propositions

It is time to raise the question again, what is required to sustain the distinction between proposition (a) and proposition (b)?

(a) ©φ
(b) (¬©)¬(©φ)

Recall that proposition (b) was found to be logically equivalent to proposition (b").

(b") ¬((¬©)φ)

The apparent conclusion is that my elaborate analysis has failed to find a way of ascribing different truth-values to the same proposition without a contradiction. This is because proposition (b") does not include proposition ©φ at all. The complex conceptual structure that I have devised seems to collapse at a crucial point—that of preserving proposition ©φ in the scope of a context operator. The problem seems to be that this containment is so potent that the proposition ©φ is absorbed in the complex conceptual structure in which I have tried to contain it. This is another aspect of the problem that I have discussed in Sect. 4.6, regarding the truth-value of ©φ while within the scope of (¬©). The containment is so powerful that it makes it difficult to refer to ©φ as an independent proposition.

It seems that any attempt to explain the ability to both affirm and deny the same proposition without a contradiction confronts a dilemma: of either securing the identity and independence of the said proposition, thus securing the ability to both affirm and deny the same proposition, but facing a contradiction; or enclosing the said proposition in the confines of a conceptual containment apparatus, which will avoid the contradiction, but ipso facto absorbing it into the structure that contains it, which thus prevents singling it out for affirmation or denial.

It is impossible to solve the problem by attempting to prevent the transition from proposition (b) to proposition (b"). The logical equivalence between the two propositions, which absorbs proposition ©φ and prevents the ascription of truth-value to it, is a necessary feature of the conceptual apparatus that was developed to allow ascribing different truth-values to the same proposition without a contradiction. To solve the problem, there must be a way of isolating and directly referring to the proposition ©φ and its truth-value within the scope of (¬©), while maintaining the complex conceptual structure that was developed here.

To do this, it is helpful to pay attention to how I have described in natural language what is required to solve this problem: What I need to do is to single out a specific proposition, that is, ©φ, and ascribe the truth-value "false" to this proposition. In formulating this end, I have relied on the conceptual resources of natural language, which enables me to speak about propositions and their truth-value. Thus, instead of simply claiming that *p*, natural language allows me to talk about the proposition *p*, and say it is true. Although a proposition *p* is logically implied by the proposition "it is true that '*p*'," there is nevertheless a fundamental difference between the two propositions because the latter proposition refers to a proposition as an entity, and ascribes to it a property, that is, a truth-value.

The possibility of speaking of the truth-values of propositions requires two conceptual abilities, which are now found to be necessary for my task. The first is the ability to refer not only to the world, but also to the propositions that describe the world. The second is the ability to ascribe to these propositions the properties of being "true" and being "false." Turning back to proposition (b), what needs to be done is to reformulate it, in such a way that it would not only use the proposition (©φ), but further refer to it as a proposition that is either affirmed or denied.

The objection may be raised that ¬(©φ) is logically equivalent to the proposition "it is true that '¬(©φ)'," and hence there is no real distinction between the two propositions. I agree that if the two propositions are logically equivalent then ascribing a truth-value to a proposition would be meaningless. This follows from the conditions for the meaningfulness of expressions in language that were revealed in Chap. 2. If the proposition ¬(©φ) is logically equivalent to the proposition "it is true that '¬(©φ)'," then the expression "it is true that '...'"fails to contribute to the determination of the truth-value of the proposition that is expressed, which proves that its appearance is meaningless. However, I do not agree that the two propositions are logically equivalent. In fact, this can be seen through a reductio ad absurdum of this objection. For, if they were equivalent, it would have been impossible to formulate the distinction between a proposition and what it describes—a distinction which is used to formulate this objection.

There is an obvious logical connection between the propositions "¬(©φ)" and "it is true that '¬(©φ)'," in the sense that the latter proposition implies the former. This logical connection is important, for it reflects the connection between the truth-value of a proposition and what it is supposed to describe. However, it is also important to acknowledge the fact that the two propositions describe different entities, and hence are not logically equivalent. What maintains the distinction between the two propositions is the fact that "¬(©φ)" does not imply "it is true that '¬©φ'." For while the former describes how things are, the latter describes the proposition that is describing how things are. The existence of a proposition does not follow from the fact that things are as they are described by this proposition.

In summary, there must be a conceptual apparatus that allows us to explicitly ascribe a truth-value to a proposition, to enable ascribing different truth-values to the same proposition. This in turn requires the distinction between the way things are and the proposition that is used to describe them.

The question that arises now is, what is needed to sustain the distinction between a proposition and what it describes? In terms of the current analysis, this is translated as the question: What are the necessary conditions for *not* being able to infer that "it is true that '¬(©φ)'," from the premise that "¬(©φ)"? This would ensure that the two propositions are not logically equivalent, and thus allow giving meaning to the identification of propositions and for ascribing to them a truth-value.

Explaining how it is possible that ¬(©φ) is true while is false that "it is true that '¬(©φ)'" would require a detailed analysis that is undertaken in the next sections. However, one necessary condition for explaining this is identifying a proposition as a concrete entity in the world, which is part of a conceptual scheme, rather than as an abstract entity, and a truth-value as an attribute of this concrete entity. Thus, this distinction requires identifying propositions as part of nature, as well as the conceptual scheme in which they are constituted, and truth-values as properties that reflect the relation between the propositions and the world. Ascribing a truth-value to a proposition hence becomes just another fact of nature. In light of the conclusion of Chap. 3, this implies that the existence of a conceptual scheme, as well as the existence of the propositions that are formulated within its confines, are contingent.

In passing, I should note that this idea not only coheres with but also gives substance to the idea of a context operator. So far, I have only characterized the use of the context operator as required for explaining contingency in language. The question of how the context operator makes it possible for the same proposition to be both true and false, not as a question about a conceptual apparatus but rather about the actual "mechanism" that makes it possible, was left aside, to be answered when the idea of a context operator is given further substance. A beginning of an answer to this question is now forming. For, by identifying a proposition as part of nature, a proposition, as a description of what exists, is given a "context"—just as a proposition is located in the scope of a context operator, it is also located in nature.

Turning back to my task, of explaining what is involved in making the distinction between a proposition and what is described by this proposition, it is clear that the ascription of a truth-value to the proposition ©φ should be formulated in terms of the conceptual scheme that is under discussion. Hence, a suitable conceptual apparatus ought to be introduced into this conceptual scheme, to enable it to refer to its own propositions and ascribe to them a truth-value.

To allow ascribing different truth-values to the same propositions, propositions (a) and (b) ought to be formulated more accurately, by explicitly ascribing truth-value to a proposition. To enable this, this conceptual scheme must be introduced with the ability to refer to propositions, which will be done with the help of a single quotation mark, and recognize truth-values as properties of propositions, which will be represented by "T" for "true" and "(¬T)" for "false." Notice that "¬T," without parenthesis is simply a negation of "true."

The conceptual characteristics of the properties "T" and "(¬T)" are to be clarified in what follows. For now, it suffices to emphasize that they should not be understood as truth-functional connectives. That is, the truth-value of the compound proposition T'*p*' is not a function of the truth-value of its sub-proposition *p*, at least in the context of (¬©). For if this were the case, the proposition (¬T)'*p*' would be equivalent to $\neg p$; while T'p' would be equivalent to p. This would render the whole attempt to speak of propositions and their truth-value meaningless. It is important to keep this in mind in the following analysis, because it highlights the fact that the ability to include truth-values in the vocabulary of a language requires an explanation. The following investigation will have to shed light on this issue.

Thus, propositions (a) and (b) are replaced with the more accurately formulated propositions (a1) and (b1).

(a1) T'©φ'
(b1) (¬©)(T'¬(©φ)')

Notice that rather than ascribing the truth-value "false" to the proposition ©φ, I have chosen to ascribe the truth-value "true" to the proposition ¬(©φ). This simplifies the following discussion because it does not require defining the logical relations between "true" and "false." Thus, strictly speaking, propositions (a1) and (b1) do not ascribe conflicting truth-values to the same proposition, but rather ascribe the truth-value "true" to a proposition and its denial. This does not affect the following discussion since the conceptual apparatus that is developed here, which enables containment of this contradiction, can thereby explain the contingency of propositions in language.

Finally, to avoid a contradiction, proposition (b1) should not imply proposition (c1).

(c1) T'¬(©φ)'

This is because propositions (a1) and (c1) contradict one another.

Notice that proposition (b1) explicitly claims that the proposition ¬(©φ) is true in the context of (¬©), and hence it is not logically equivalent to proposition (b), that is, (¬©)¬(©φ), and hence is also not equivalent to proposition (b"), that is, ¬((¬©) φ). Proposition (b1) implies proposition (b), and this is important for not undermining the meaning of "true." However, proposition (b) does not imply proposition (b1). Thus, the problem that was raised at the beginning of this section is solved. Propositions (a1) and (b1) necessitate both affirming and denying the same proposition, thus explaining the need for contingent propositions in describing what exists.

However, it is not enough to stipulate that proposition (b) does not imply proposition (b1). This fact is supposed to reflect the conceptual connections that obtain, or do not obtain, in the conceptual scheme in which these propositions are formulated. Hence, this conceptual scheme must include the possibility of proposition (b) being true while proposition (b1) is false. That is, it must acknowledge that proposition (d) describes a conceptual possibility.

(d) (¬©)¬(©φ) ∧ ¬(¬©)(T'¬(©φ)')

Before concluding this section, I should address the question of the relation between proposition (a1), that is, T'©φ', and proposition (a), that is, ©φ. Again, due to the meaning of a truth-value, proposition (a1) must imply proposition (a).

However, is there a need to maintain a distinction between these propositions, such that proposition (a) does not imply proposition (a1)?

To begin with, I should emphasize that this is not implied by the previous analysis. To give meaning to the ascription of truth-values to propositions, it suffices that is meaningful in the scope of the context operator (¬©). Furthermore, to avoid a contradiction, only one proposition should be avoided, and my analysis led me to the conclusion that the proposition that is in the scope of the context operator (¬©) is the one that should be avoided. This fact is not coincidental. The context operator © was determined to be the default context operator, and hence this context operator could not provide the possibility of containing the proposition ©φ.

The last fact also implies that it is impossible to make the distinction between the truth of ©φ and the ascription of the truth-value "true" to this proposition. At this stage of the analysis, it is clear that propositions that include the context operators are not abstract entities, but are rather concrete entities that are part of a conceptual scheme and, in fact, ought to be recognized by the conceptual scheme as such. This implies that it is impossible, in the confines of a conceptual scheme, to affirm proposition ©φ without ipso facto identifying this proposition as a proposition in the conceptual scheme. Hence a conceptual scheme cannot affirm this proposition without ipso facto ascribing to this proposition the truth-value "true" and accepting the proposition T'©φ' as true.

Notice that the same is the case regarding proposition (b1), that is, (¬©) (T'¬(©φ)'). If this proposition is affirmed in a conceptual scheme, in this conceptual scheme proposition (b1') must also be affirmed.

(b1') T'(¬©)(T'¬(©φ)')')

To reiterate, the reason that the proposition ¬(©φ), in its appearance in proposition (b), is not replaceable with the proposition T'¬(©φ)' is that it is within the scope of the context operator (¬©), which is not the default context operator. In this respect, the context operator (¬©) is similar to a propositional attitude. Likewise, "I believe that ¬(©φ)" does not imply that "I believe that T'¬©φ'." One can believe a proposition without believing anything that is implied by it, and this means that propositions that are logically equivalent in some contexts may not be logically equivalent in other contexts. Again, this is due to the unique function of the context operator.

However, it is not enough to argue that this is how the context operator (¬©) functions. It is necessary to explain what makes this unique function possible. Indeed, the reader may wonder how this can be possible, given the conceptual connection between the ideas of a proposition and a truth-value. The answer to this query will be found in what follows, by uncovering the necessary conditions for this unique function of the context operator (¬©).

4.10 Context Operators as Temporal Indexical Expressions

So far, I have attempted to conduct my investigation as independently as possible from our conceptual scheme. I have relied on temporal indexical expressions in natural language only as an analogy to help the reader understand the implications of the abstract conceptual structure I have been developing in this chapter. At this point, however, I have established a sufficiently detailed conceptual structure to identify the context operators with temporal indexical expression. For this purpose, I shall rely on two claims. The first is that in ordinary language it is temporal indexical expressions that enable ascribing different truth-values to a proposition without a contradiction. All other expressions in ordinary language either do not enable this or rely on temporal indexical expressions. The second is that the conceptual structure of temporal indexical expressions is identical to the structure of the context operator. Taken together, these claims imply that time is a necessary feature of nature.

To begin with, a distinction should be made between demonstrative expressions and indexical expressions (Kaplan, 1989, pp. 490–491). Examples of demonstrative expressions are "this," "that," "he," "she," and similar expressions; while examples of indexical expressions are "I," "here," "now," and similar expressions. The difference between demonstrative and indexical expressions is that the reference of demonstrative expressions is determined, inter alia, by something that accompanies their utterance, for example, an accompanying act of the speaker, or simply by the intention of the speaker. This means that it is impossible to simply utter these expressions, without an accompanying act of demonstration and successfully express a proposition. The reference of indexical expressions, on the other hand, does not depend on any accompanying act. It is rather determined by the circumstances of their use.

Sentences that contain demonstrative or indexical expressions express propositions that can be both true and false in different circumstances. However, the question is whether these sentences enable the expression of the same proposition in these circumstances. Only if the same proposition is both true and false in different circumstances can the contingency of propositions in language be explained.

To determine whether the same proposition is expressed in different uses of a sentence, I shall rely on the following consideration. If distinct propositions are expressed by various utterances of the same sentence in different circumstances, it should be possible to express the same propositions by the use of different sentences. This is because of the distinction between the sentence, as a combination of arbitrary signs, and the meaning of the signs, that is, their function, which constitutes the proposition. To claim that certain signs express different propositions in different circumstances implies that the same sign is used in different ways in different circumstances. Hence, it should be possible in principle to distinguish between the different uses by the use of different signs.

Hence, the criterion for determining whether the same or different propositions are expressed in different utterances of a sentence is this: different propositions are expressed in different circumstances by the same sentence if and only if it is possible to express the same propositions by the use of different sentences. I should note that I do not assume that every language de facto includes sufficient vocabulary to express different propositions by the use of different signs. I do assume, however, that it is possible to introduce different signs to express the different propositions. That is, I assume that it is possible to formulate a language that will clearly reflect the conceptual structure of the propositions it is able to express.

A clarifying note is in order at this point. The claim I am making here is well known, as it is at the heart of contemporary debates concerning the nature of time. Specifically, the claim that the same proposition can be both true and false in different contexts directly challenges "semantic eternalism," the view that propositions are eternally true or false, regardless of context, and defends "sematic temporalism" (Brogaard, 2012). However, I emphasize again that I am not committing to any current view in the philosophy of time, nor do I assume this as a premise of my analysis. Rather, I have established this as a requirement for the contingency of propositions in language, and in what follows, I seek to demonstrate that natural language indeed allows for this. Furthermore, I believe that the ongoing debate on this subject is unresolvable because it is conducted without the foundation of established principles of conceptual analysis—an unfortunate feature of contemporary philosophy that this book seeks to address.

Hence, I must stress that my argument relies on the concept of language as a system of coordinates. What I claim is that if different propositions are expressed in different circumstances by the same sentence, they thereby denote different conceptual possibilities in conceptual space. Consequently, it must be possible to distinguish between these possibilities using different coordinates. To be clear, I do not argue that if different propositions are expressed in different contexts, different sentences must be used. This is obviously a false claim. What I argue is that if different propositions are involved, different sentences could be used to express them (without loss of meaning).This leads me to question whether demonstrative and indexical expressions can be reduced to expressions that do not include any expression of these categories. If such a reduction is possible for any such expression, then this expression cannot explain the possibility of ascribing different truth-values to the same proposition without a contradiction. For this would imply that this expression does not enable the expression of different propositions in different circumstances.

However, this criterion implies that if a demonstrative or an indexical expression is irreducible to an expression that does not include demonstrative or indexical expressions, then it is impossible to express the propositions it expresses in different contexts by the use of different sentences. This in turn implies that it expresses that same proposition in each context of its use.

To begin with, demonstrative expressions are reducible, and hence do not express the same proposition in different contexts of their use. The ability to reduce these expressions to expressions that do not include demonstrative expressions is explained by the possibility of replacing each individual occurrence of these expressions with a description of the specific accompanying act of demonstration that accompanies it or the intention of the speaker who uses it, which determines their reference. Thus, they are in principle eliminable. If they are reducible to expressions that do not include indexical expressions, then they cannot explain the ability to ascribe different truth-values to the same proposition without a contradiction. Some of them may, however, be reducible to expressions that include indexical expressions, to which I now turn my attention.

Contrary to demonstratives, indexical expressions cannot be reduced to expressions that do not include any indexical expressions. This claim is not new. It was first made by John Perry, although he did not make the distinction between demonstrative and indexical expressions (Perry, 1977, 1979).

Perry's claim was made in terms of Frege's conception of language. According to his claim, there is no single expression with the same sense as an indexical expression, such that it can replace every occurrence of this indexical expression without a change to the meaning of the sentence. This may seem trivial given the fact that this claim is made in terms of Frege's conception of language. That is, it might seem trivial that no sentence that does not include an indexical expression, and hence expresses a proposition with a fixed truth-value, can have the same sense as a sentence that includes an indexical expression, and hence expresses in different contexts propositions with different truth-values.

However, it is further claimed that even a single occurrence of an indexical expression cannot be replaced with an expression that does not include any indexical expressions without a change in the sense, given Frege's criterion of a difference in sense (Frege, 1948, pp. 209–210). That is, for any expression φ, with which a particular occurrence of an indexical expression is replaced, for example, "now," it is possible to believe that "The meeting starts now," but not believe that "The meeting starts at φ," and vice versa.

Furthermore, even if there is a sentence of the type "The meeting starts at φ," which the subject holds as true while believing that "The meeting starts now," the reference of φ may not be identical with the reference of the indexical expression "now" (Perry, 1977, pp. 486–488, 1979, pp. 7–8). This is because the indexical expression's reference is determined by the context, and hence is independent of the subject's beliefs about the current time, which may well be wrong.

As mentioned earlier, these arguments are formulated in terms of Frege's conception of language; hence, I cannot rely on Perry's considerations. Furthermore, Perry's argumentation—especially his reliance on the special explanatory role of indexical expressions in the context of explaining actions, poses additional issues (Cappelen & Dever, 2013). Beyond the specific criticism of Perry's arguments, it is also unacceptable, within the framework of this inquiry, to rely on the complex conceptual notion of "action," which itself depends on substantial conceptual assumptions. As previously stated, I will not rely on considerations that are not based on the analysis of language conducted in Chap. 2.

However, it is possible to formulate an independent argument, expressed by viewing language as a system of coordinates, that leads to a similar conclusion. The fact that the reference of indexical expressions is independent of the subject's beliefs implies that the claim that "now is φ" is always a meaningful identification. That is, in the sentence "The meeting starts at φ, and φ is now," the second conjunct contributes to the meaning of this sentence. Hence, it is impossible to reduce any specific appearance of an indexical expression to an expression that does not include any indexical expressions.

This last implication is important, because, as our previous analysis has shown, this implies that sentences that include indexical expressions express the same proposition on each occasion of their use. Hence, indexical expressions in natural language can explain the ability to ascribe conflicting truth-values to the same proposition, and hence explain the need for contingent propositions for describing what exists.

The fact that indexical expressions in general *can* explain the ascription of different truth-values to the same proposition does not imply that every indexical expression actually enables that. Furthermore, the conclusion that indexical expressions are not reducible to expressions that do not include any indexical expressions does not imply that some indexical expressions are not reducible to other indexical expressions. If some indexical expressions are reducible to other indexical expressions, they cannot be considered necessary for explaining the need for contingent propositions in describing what exists, and hence cannot be considered to describe necessary features of nature.

Elsewhere I explored the different indexical expressions that can be found in language, to determine which can explain the ability to ascribe different truth-values to the same proposition without a contradiction, and which of those which enable this is the most basic (Yehezkel, 2014, Chap. 4). These include the first-person indexical expression, that is "I;" spatial indexical expressions, for example, "here;" and temporal indexical expressions, such as "past," "present," and "future." My conclusion was that it is the temporal indexical expressions that enable the same propositions to be both true and false in different circumstances.

However, in the present inquiry, I wish to explore a different path. This is because I have already analyzed the conceptual structure that allows ascribing different truth-values to the same proposition. What I wish to do is explore the conceptual components that are found in natural language, and examine their correspondence to this conceptual structure, which is a necessary feature of nature and a conceptual scheme. This line of inquiry has the added benefit, not only of identifying the component in language that enables a proposition to be both true and false, but of identifying the features of this component that reflect the necessary features of nature.

After introducing the context operator, that is, ©, as a hypothetical conceptual component that allows the same proposition to be both true and false, I have argued that this operator must allow for high-order uses. That is, if the basic proposition that it allows formulating is ©φ, then it must also be meaningful to apply the context operator or its negation to this proposition, and so on. These higher-order uses are demonstrated in Table 4.5.

As is evident from Table 4.5, which describes the conceptual structure that is required to support the context operator and enable it to explain the ability to ascribe different truth-values to the same proposition without a contradiction, this conceptual structure forms a linear series. This series is constituted by the help of the context operator, and another context operator that must be located on the same conceptual dimension, and which I have referred to by negation of the context operator. Hence, only the indexical expressions in natural language that enable constituting such a dimension are eligible for identification as the context operator ©.

There are only two categories of indexical expressions in natural language that enable the creation of similar linear series, the temporal and the spatial. Both categories include various indexical expressions. For example, among the spatial indexical expressions are "here, "left," "right," "north", "south," and so on. Among the temporal indexical expression are "now"/"present" (these terms will be used as interchangeable in the current inquiry), "past," "future," and similar expressions.

To begin with, it should be noticed that many of these expressions can be analyzed as relations, which can be defined with the help of other indexical expressions. For example, the expression "to the south" can be analyzed as a relation between two places and hence, if it is used, simply as an indexical expression. According to this analysis, the claim "the exit is to the south," can be analyzed as a relation to "here," that is, as short for "the exit is south of here." What is unique in the indexical expression "here" is that it refers to the exact location of the subject, in relation to which other spatial indexical expressions seem to refer, and hence ultimately rely on. Similarly, the temporal indexical expression "now" seems to be the focal point in relations, based on which other temporal indexical expressions can be defined. Thus, "here" and "now" are found to be the basic indexical operators, which enable defining all other spatial and temporal indexical expressions. Nevertheless, although this consideration is correct, I should emphasize that according to the conception of language as a system of coordinates, as well as the analysis of the previous sections, there must be at least another basic indexical expression on the conceptual dimensions on which both these indexical expressions are located, which is represented by a negation of the context operator, that is, ¬©.

Both "here" and "now" display the same unique behavior of the context operator © as the default context operator. Recall that as the default context operator © can be added without a change to the truth-value of propositions that include context operators. That is, for any proposition p that includes a context operator, p is logically equivalent to ©p.

Analogically, the spatial indexical expression "here" can be added to the sentence "it is raining here," that is, "here it is true that 'it is raining here'," without affecting the truth-value of the proposition that is expressed. Similarly, "the exit is to the south" is equivalent to "here it is true that "here it is true that 'the exit is to the south'." This is also true regarding the temporal indexical expression "now." The sentence "it is raining now" is logically equivalent to the sentence "it is true now that 'it is raining now';" and "it was raining in the past" is logically equivalent to "it is true now that 'it was raining in the past'."

The fact that the indexical expressions "here" and "now" display the same unique characteristics as the default context operator enables me to explain more clearly the meaning of this feature. The default context operator refers to a unique context, from which nature is described. Hence, it does not contribute to the meaning of sentences in its scope that include context operators. All the descriptions in a conceptual scheme are formulated from the perspective of the default context operator.

The question that remains is: Which of the indexical expressions, the temporal or the spatial, is more basic, if either? That is, the question is whether it is possible to reduce one to the other, but not vice versa. If this is not the case, both expressions ought to be recognized as equally fundamental.

An analysis of the function of both expressions in natural language shows that the spatial indexical expression "here" relies on the temporal indexical expression "now," but the reverse does not hold. To begin with, notice that the spatial indexical "here" allows for the same proposition to be both true and false, thus explaining the need for contingent propositions for describing what exists, but this is only possible due to changes in time. Thus, it can be both true and false that "it is raining here," but only if this proposition relates to different times.

It might be argued that it is nevertheless possible to explain the possibility of expressing both true and false propositions with the help of the indexical expression "here" without relying on a change in time. For example, it has been suggested that one can simultaneously utter two different tokens of the expression "here," which refer to different locations, by holding up cards on each hand with "here" inscribed on them (Gale, 1969, p. 407). However, this suggestion relies on the use of the expression "here" as a demonstrative, because it relies on an act of demonstrations, rather than as an indexical expression, which refers exclusively to the spatial location of the subject who uses this expression.

Another suggestion returns to the idea of two different subjects, in two different locations, who say "it is raining here," to explain the possibility of expressing both true and false propositions at the same time. However, it is important to notice in this context that it is necessary for both occurrences of the proposition "it is raining here" to be part of the same conceptual scheme. Hence the possibility of ascribing each of the occurrences of this proposition to a different subject fails to explain why a conceptual scheme needs to include contingent propositions to describe what exists. Although the two subjects may be using the same type of conceptual scheme, they are nevertheless numerically different. Hence this cannot explain why a conceptual scheme needs contingent propositions for describing nature.

It might be thought that describing this possibility, from the point of view of either of the two subjects, nevertheless requires ascribing different truth-values to the same proposition. However, ascribing one of these occurrences to another subject renders the indexical expression reducible. For example, "for Jack it is false that 'it is raining here'," is logically equivalent to "it is not raining where Jack is located." To explain how the same proposition can be ascribed different truth-values, it is therefore necessary to remain within the confines of a single conceptual scheme, that is, from the point of view of a single subject.

Notice, however, that the reverse does not hold. There is no need to explain the possibility of ascribing different truth-values to a proposition that includes the temporal indexical expression "now," to rely on a change in spatial position. The proposition "it is raining now" can be both true and false at different times, and in the same conceptual scheme, even if the location is the same.

The fact that the ability to ascribe different truth-values to the same proposition, using the spatial indexical, is only possible if a change in time is assumed, while the reverse does not hold, proves that time is more fundamental than space. Furthermore, this is also reflected by the fact that the spatial indexical expression "here" is reducible to the temporal indexical expression "now," while the reverse does not hold. "Here" can be defined as "(my) spatial location now," while "now" or "present" cannot be defined as "(my) temporal location here." This is because I can be at the same place at different times, but at each instant in time I am located in one place (Dolev, 2000, p. 70; Hardin, 1984, p. 122; Mellor, 1998, pp. 95–96).

In passing, I shall remark that this analogy between time and space is not accidental. However, discovering the connection between time and space and understanding their deep connection, which also explains their structural affinity, will only be explored in Chap. 6 of this book.

In conclusion, the context operator ©, which I have constructed to explain the possibility of a proposition being both true and false, is identical to the temporal indexical expression "now," or "present," in the natural language. Not only does it function similarly to the context operator, in allowing a proposition to be both true and false in different contexts, that is, at different times, but it also functions as the default context operator, and is a basic and irreducible feature of our conceptual scheme. The basic propositions in language are hence propositions that include the temporal indexical expression "now." The temporal location from which nature is described thus explains the need for contingent propositions for describing what exists. Existence is hence determined in relation to an internal point of view of nature. That is, existence is a relation between constituents of nature. The implication of this conclusion will be explored in Chap. 7 of this book.

The current investigation, however, has not concluded. In natural language, the indexical expression "now," or "present," is located on a conceptual dimension that includes two more indexical expressions, "past" and "future." Recall that the justification for introducing the context operator ¬© was the fact that a minimal conceptual dimension includes two possibilities. However, I should emphasize that I did not presume in the previous analysis, by introducing another context operator (a negation of ©), that these are the only two context operators located on this conceptual dimension. The negation sign was only meant to signify the fact that it is not the context operator ©.

While the previous analysis only found the need for two context operators, the fact remains that in natural language there are three indexical expressions on the conceptual dimension that includes the indexical expression "now." Two questions now arise. The first is: How can the existence of a third indexical expression be explained? The second question is: Should the second context operator that was identified during my analysis, that is, ¬©, be identified as "past" or as "future" in natural language?

The two questions are intertwined, and I believe that answering the first question will provide a guide for answering the second. The fact that my analysis has only found the need for two basic context operators suggests that the third indexical expression is derived from the other two. This leads me to the conclusion that the context operator ¬© should be identified with the temporal indexical expression "past." I shall explain and justify this identification more thoroughly in Chap. 5. Meanwhile, I shall explain that the reason for this identification is that, in our conceptual scheme, the use of the temporal indexical expressions "now" and "past" is more basic than the use of "future," in the sense that they are used in direct descriptions of what is available from the point of view of the conceptual scheme. The use of the indexical expression "future," on the other hand, seems to be restricted to inferential uses, that is, to conclusions that we draw based on what we directly experience. In this sense, the use of the temporal indexical expression "future" relies on the more basic use of "now" and "past." Hence, I identify the context operator ¬© with the temporal indexical expression "past" in natural language.

An objection to my claim that "past" is a basic concept in our conceptual scheme may be based on the claim that it can be defined with the help of the temporal indexical "now" and the temporal relation "before," simply as "before the future." In answer to this objection, it should first be noted that my analysis proves that the concepts of present and past are interdependent—recall in this context that a minimal conceptual dimension must include at least two distinct conceptual possibilities. Hence it is impossible to begin by independently giving meaning to the indexical expression "now," and only later introduce the indexical expression of "past." They are hence both basic concepts in any conceptual scheme. Furthermore, the fact that they are basic means that other concepts, including the temporal relations of before and after, should be defined using these distinctions. Indeed, at this point of the investigation, I have yet to give meaning to these temporal relations, and hence they are not available as part of the conceptual apparatus that is available at this stage of my analysis.

I should emphasize that in these identifications, of the theoretical context operators with the temporal indexical expressions in natural language, I do not commit to the full conceptual apparatus of our conceptual scheme. In what follows I rely only on the conceptual structure that I have developed in this chapter. Introducing any further conceptual apparatus must rely on this structure, and cannot be justified based on our existing conceptual scheme.

It is now time to introduce a new notation. I replace the context operator © with the capital letter N (for "now"), the context operator ¬© with the capital letter P (for "past"), and introduce the capital letter F for the future.

Replacing our previous notation with the new interpretation gives us new insight into the nature of the complex conceptual structure that was devised during my analysis. According to this interpretation, Table 4.5 is now translated to Table 4.6.

Table 4.6 Temporal interpretation of Table 4.5

4.6.1	Nφ
4.6.2	Pφ
4.6.3	P(Pφ)
4.6.4	P(P(Pφ))
	…

The temporal conceptual series that is depicted in Table 4.6 begins with proposition (4.6.1), which describes what exists now. Proposition (4.6.2) describes what existed in the past. It is however a general proposition that claims that somewhere in the past the property φ was realized. Proposition (4.6.3) claims that in relation to the past, the property φ was realized in the past, and so on. However, notice that this does not constitute a temporal series of occurrences from the near past to the distant past. This is because lower-order propositions, such as proposition (4.6.2), do not describe the near past, but are rather general propositions that claim that somewhere in the past φ was realized. Hence, proposition (4.6.2) follows from any higher-order proposition, that is, (4.6.3) and higher. The question of how this complex structure enables us to constitute time as a linear dimension of occurrences will have to be explained in what follows, and will demonstrate how the notion of the future can be introduced (in Sect. 4.13).

Lastly, I would like to demonstrate how time allows the same proposition to be both true and false. Recall that my search for the possibility of the same proposition being both true and false led me to the following formulations:

(a1) T'©φ'
(b1) (¬©)(T'¬(©φ)')

These propositions are now identified with propositions (a2) and (b2) respectively in our conceptual scheme.

(a2) T'Nφ'
(b2) P(T'¬(Nφ)'))

Proposition (a2) claims that now the proposition "now φ" is true. Proposition (b2) on the other hand, claims that in the past its denial was true. Thus, to describe this change, the proposition "now φ" must be a contingent proposition. The change in the determination of present, that is, the flow of time, is hence the property of nature that requires the use of contingent propositions for its description. It is therefore a necessary and fundamental feature of nature.

The conclusion of my analysis is that time, as it is described by the distinctions between past and present (now), is a necessary feature of nature, which explains the need for contingent propositions for describing what exists. Furthermore, these temporal distinctions describe nature from the point of view of a conceptual scheme which, to give meaning to these distinctions, must describe nature from a specific temporal location, that is, the present, in relation to which the more general notions of the past and the future are determined. This point of view of nature changes, however, in time. This change in position explains the need for contingent propositions for describing what exists.

4.11 Objectivity I: Distinguishing the Conceptual Scheme and the World

In the previous section, I identified the context operators with the temporal indexical expressions "present," or "now," and "past." Propositions (a1) and (b1), which describe how the same proposition is affirmed and denied, were hence translated as propositions (a1) and (b2).

(a2) T'Nφ'
(b2) P(T'¬(Nφ)'))

However, there is a concern that was addressed briefly in Sect. 4.9 and needs to be addressed now. Recall that proposition (b2) is supposed to be distinguished from proposition (b), which is now translated as proposition (b3).

(b3) P(¬(Nφ))

Proposition (b3) is logically equivalent to proposition (b4).

(b4) ¬Pφ

Notice that proposition (b4) does not include denial of proposition Nφ. It simply states that in the past it was not the case that φ was realized. Hence, proposition (b3) cannot explain the need for contingent propositions for describing what exists. What is described by proposition (b3) can also be described without denial of proposition Nφ.

Proposition (b2) can hence explain the need for contingent propositions for describing nature only if it is not logically equivalent to proposition (b4). The difference between propositions (b2) and (b3) is that, while in proposition (b2) it is claimed that in the past it was true that "it is not that case that now φ," in proposition (b3) it is claimed that in the past "it is not that case that now φ." I have argued that the notion of truth implies that proposition (b3) follows from proposition (b2). That is, *p* follows from T'*p*.'

Hence, the only way to sustain the distinction between propositions (b2) and (b3), and support the meaningfulness of the ascription of truth-value in proposition (b2), is if proposition (b3) does not imply proposition (b2). This in turn requires that the conceptual scheme in which these propositions are formulated includes the conceptual resources that are needed to acknowledge the possibility that is described by proposition (d1).

(d1) $P(\neg(N\varphi)) \wedge \neg P(T`\neg(N\varphi)')$

To allow for the conceptual possibility that is described in proposition (d1), a distinction must first be made between a relative and an actual truth-value. To understand this distinction, let me begin with a relative truth-value. According to this idea, when a claim is made that under certain conditions, or in a certain context, a proposition is true, all that it implies is that if this proposition would have been asserted under these conditions, this proposition would have been true. Thus, for example, it can be said that at the time of the dinosaurs, the proposition "there are dinosaurs now" is true. According to the relative interpretation of this claim, what is claimed is that in relation to the time of the dinosaur's existence, this proposition is true. This statement is equivalent to the claim that "there were dinosaurs in the past," and is not committed to anything beyond that straightforward claim.

There are, however, cases in which it is wished to describe an actual truth-value of a proposition. This is the sense in which it is correct to say, for example, that the proposition "there are dinosaurs now" has never been true. This is because no one who possesses a conceptual scheme that includes sufficient conceptual resources to formulate this proposition lived at the time of the dinosaurs. Hence this proposition was never actually true. This is why the claim that the proposition "'there are dinosaurs now' was actually true in the past" says more than that "there were dinosaurs in the past," and hence these claims are not logically equivalent, although the former implies the latter.

Notice that the distinction between a relative and an actual ascription of truth-values does not apply to every proposition. It is possible to make this distinction only about propositions that are sensitive to the context of their use, for this context is the reference point in relation to which they are ascribed with a truth-value. Interestingly, due to grammatical tense, this sensitivity to a context permeates almost all the propositions in language that are used to describe what exists—a fact that finds its explanation in the current inquiry. Thus, for example, even the proposition "dinosaurs roamed the earth for over 165 million years," is formulated in the past tense, and hence was not actually true at the time of the dinosaurs. However, it is possible to stipulate an atemporal use of the present tense, for example, which is not sensitive to the context of its use. Thus, according to this convention, the proposition "dinosaurs roam the earth for over 165 million years" is true regardless of the time of its use, and hence it is impossible to apply the distinction between a relative and actual truth-value to this proposition. I shall show in Sect. 4.13 how this use of language can be accurately defined using the conceptual tools that are developed here.

In passing, I shall note that the failure to distinguish between the relative and the actual truth-value is important for describing the "flow of time," that is, the change in the distinctions between past, present, and future. Take, for example, the description "occurrence e was future, is present, and will be past." If this description is interpreted in relative terms, it fails to capture the flow of time, as evident from the fact that as a relative statement, it has a spatial analogy, that is, "occurrence *e* is north in the south, present here and south in the north", which is equally true (Taylor,

1955, pp. 607–608). Only as a description of the actual change in the truth-values of propositions that include temporal indexical expressions can the flow of time be properly captured.

Turning back to proposition (d1), that is, $P(\neg(N\varphi)) \wedge \neg P(T`\neg(N\varphi)')$, in applying the distinction between a relative and an actual truth-value to the conceptual possibility that is described by this proposition, there are several suggestions as to how this can be explained. In surveying them, it is important to remember that the distinction between the relative and the actual interpretation relies on the presumption that propositions and the conceptual scheme in which they are formulated are part of nature. Furthermore, remember that the conceptual resources that are available at this stage of the inquiry allow the conceptual scheme to relate to its propositions and their truth-values.

I begin by exploring three alleged explanations for the truth of the second conjunct in proposition (d1), that is, $\neg P(T`\neg(N\varphi)')$.

(1) In the past, the conceptual scheme and hence the conceptual resources for formulating the proposition $\neg(N\varphi)$ existed. However, this proposition was not actually true, but rather was false.
(2) In the past, the conceptual scheme and hence the conceptual resources for formulating the proposition $\neg(N\varphi)$ existed. This proposition was true at that time, but it was not recognized as true in the conceptual scheme. That is, it was either (mistakenly) considered false, or its truth-value was not decided.
(3) In the past, the conceptual scheme did not exist at any time in which the property φ was not realized. Hence, the proposition $\neg(N\varphi)$ was never actually true in this conceptual scheme.

To begin with, suggestion (1) should be disqualified as a valid explanation for the truth of proposition (d1). For it contradicts the first conjunct of proposition (d1), according to which the proposition $\neg(N\varphi)$ was true in the past.

Suggestion (2) relies on the possibility of making the distinction between the actual truth-value of a proposition and the truth-value that was recognized or ascribed to this proposition. In referring to our conceptual scheme, this possibility is demonstrated by someone who either does not know whether the earth revolves around the sun, or by someone who mistakenly thinks that the earth does not revolve around the sun. The first possibility I term "the possibility of ignorance," and the second possibility I term "the possibility of mistake."

I should emphasize at this point that although the proposition $P(T`\neg(N\varphi)')$ ascribes the truth-value "true" to the proposition $\neg(N\varphi)$, it does not imply anything about whether the proposition $\neg(N\varphi)$ was considered to be true in the past in any conceptual scheme. It claims that this proposition was actually true in the past. To understand the distinction I am making here, think of the following example. When I say that the proposition "the earth revolves around the sun" was actually true in the Middle Ages I am not claiming that anyone actually thought it was true (although there may have been scholars in the Middle Ages who believed it). My claim is about the actual truth-value of this proposition, rather than the truth-value that it was thought to have. Similarly, proposition $P(T`\neg(N\varphi)')$ must be understood as a claim about the actual truth-value of the proposition $\neg(N\varphi)$, rather than what it was considered to be. Hence, its denial, that is, the second conjunct of proposition (d1), denies that it was actually true in the past.

The reason for this formulation of the second conjunct of proposition (d1) is that I am trying to explain the need for contingent propositions in language, and my analysis concluded that it is only possible to explain this need if it is necessary to use propositions that are both actually true and actually false to describe what exists. Thus, for example, merely considered, perhaps by mistake, as both true and false is not sufficient for explaining the need for contingent propositions, because even necessary propositions can be mistakenly considered to be false. Furthermore, I have yet to find it necessary for this proposition to be *acknowledged* as both true and false, to explain the need for contingent propositions for describing what exists. All that is required is that it is actually both true and false at different times.

Hence, suggestion (2) does not explain the truth of proposition (d1). It asserts that the proposition $\neg(N\varphi)$ was not held to be true in this conceptual scheme. This, however, is not what is claimed in proposition (d1). Proposition (d1) stipulates that the proposition $\neg(N\varphi)$ was not true as a matter of fact.

Suggestion (3) expresses the only way to explain the possibility that is described by proposition (d1). According to this suggestion, the first conjunct in proposition (d1), that is, $P(\neg(N\varphi))$, is interpreted as a relative statement, according to which proposition $\neg(N\varphi)$ is true in relation to the past. That is, in the past φ was not realized. The second conjunct, on the other hand, denies that the proposition $\neg(N\varphi)$ was actually true in the past. This can only be if this proposition was not available for describing nature at that time.

Hence, to explain the possibility of proposition (b3) not implying proposition (b2), the ascription of the truth-value "true" to proposition $\neg(N\varphi)$ in proposition (b2) must be actual, rather than relative. This requires understanding the proposition $\neg(N\varphi)$ not as an abstract entity, but rather as an existing entity in the world, as part of a conceptual scheme that is used to describe nature. This is the only way in which, notwithstanding the fact that in the past φ was never realized, the claim that $\neg(N\varphi)$ was never actually true (in the confines of this conceptual scheme).

Notice that suggestions (2) and (3) imply objectivity, although in two different senses. Suggestion (2) distinguishes between what exists and the conceptual scheme in which it is described. It assumes that what exists can be different from what is accepted as true in a conceptual scheme. However, as I explained, it does not directly explain the possibility that is described by proposition (d1).

Suggestion (3), on the other hand, relies on a more robust notion of objectivity. For, according to this suggestion, the proposition $\neg(N\varphi)$ was not actually true in the past, although it would have been true had it been made. This is only possible if the conceptual scheme did not exist in the past and hence was not available for describing nature.

To understand how far-reaching this suggestion is, let me again use an example from our conceptual scheme. Thus, I can truly say that although in the Mesozoic Era the proposition "there are dinosaurs now" was true in the relative sense, that is, in the sense that if it was made it would have been true, it was never actually true because no human being existed at that time.

Thus, suggestion (3) not only requires viewing propositions as entities that exist (or not) in the world, but it also requires viewing the conceptual scheme as an entity that exists in the world at some times, but not at others. It describes nature from a limited point of view that is internal to nature.

In conclusion, explaining the contingency of propositions requires relying on suggestion (3), and thus also acknowledging the non-existence, and hence also the existence, of the conceptual scheme at certain times. This also applies to propositions in language, that must be understood not only as abstract entities but rather as entities that exist in time. Again, this is expected with regard to propositions that are sensitive to the time of their use, that is, which include temporal indexical expressions. The temporal indexical expressions reflect the point of view from which nature is described, and hence characterize these propositions, which thereby acquire a location in time. Furthermore, this inquiry shows that this characterization also applies to the conceptual scheme itself, which describes nature from a point of view in time, that is, the present, which forms the basis for any description of nature.

4.12 Objectivity II: Identifying Propositions and Truth-Values

In the previous section, I argued that only suggestion (3), which implies that the proposition $\neg(N\varphi)$ was never actually true in the past, although if it was made in this context it would have been true, can explain the conceptual possibility that is described by proposition (d1). Suggestion (2) does not explain the truth of proposition (d1), because it merely asserts that the proposition $\neg(N\varphi)$ was not acknowledged as true in this conceptual scheme, although it was in fact true.

However, although suggestion (2) fails to provide sufficient conditions for supporting the conceptual possibility that is described by proposition (d1), it is nevertheless a necessary condition for including this conceptual possibility in any conceptual scheme. To see why, I wish to explore more carefully the conceptual structure that is reflected in proposition (d1).

(d1) $P(\neg(N\varphi)) \wedge \neg P(T`\neg(N\varphi)')$

My analysis has led me to the conclusion that, to explain the need for contingent propositions for describing nature, propositions should not be thought of as abstract entities, but as concrete entities that exist as part of nature. This implies that, while it is possible to sometimes talk of a proposition in abstraction from the signs that are used as coordinates in a conceptual scheme, the most basic propositions in language must be thought of as concrete uses of signs, that is, as a token rather than a type. More specifically, the second conjunct in proposition (d1) refers to the fact that in the past the proposition $\neg(N\varphi)$ was not actually available.

As mentioned before, the conceptual possibility that is described by proposition (d1) must be included in a conceptual scheme, for this conceptual scheme to enable the formulation of contingent propositions. Because this possibility requires an ability to refer to propositions and acknowledge their existence as part of nature, this conceptual scheme must be able to identify a proposition as a proposition.

This fact implies that the conceptual scheme must be able to make the distinction between a proposition and what this proposition describes in the world. Only a conceptual scheme that recognizes the existence of propositions could have the conceptual resources to formulate proposition (d1), in which proposition ©φ is not only used, but also referred to with the help of single quotation marks, that is, with the expression '¬(Nφ).'

It is not easy, however, to explain how it is possible to recognize the existence of a proposition in the confines of the conceptual scheme in which it is constituted. For in this conceptual scheme, when the combination of signs ¬(Nφ) is used, it is first and foremost used to describe nature. In this use of the signs, the use itself is "transparent," that is, it is not reflected in the meaning of these signs. That is, the signs describe nature, rather than their description of nature. The question is, how can a conceptual scheme acknowledge its propositions as such, that is, as descriptions of nature, in distinction from what they are used to describe?

The fact that the basic proposition in language that is referred to, that is, ¬(Nφ), is not thought of as an abstract entity, but rather as a specific use of signs in a particular context, enables making this distinction. For it is simply a distinction between two separate entities in the world, and as such, this proposition is subject to the necessary conditions of any existent, that is, it is conceived as a contingent entity in the world. However, although these are two distinct entities, there is also a conceptual connection between the proposition and what it describes. The problem is hence not to identify two separate entities in nature, but rather to make a distinction between two conceptually connected entities.

To include the conceptual possibility that is depicted by proposition (d1) in a conceptual scheme, this conceptual scheme must incorporate the conceptual apparatus that enables acknowledging the existence of propositions, that is, an entity to which a truth-value can be ascribed. However, the mere existence, or non-existence, of the conceptual scheme, which was found to be a necessary condition for explaining proposition (d1) in the previous section, is not sufficient for explaining the acknowledgment.

To see why acknowledging the mere existence or non-existence of a conceptual scheme is not sufficient to explain the conceptual structure that is reflected in proposition (d1), it is worth comparing it to proposition (d2).

(d2) $P(\neg N\varphi) \wedge \neg P(\text{`}\neg N\varphi\text{'})$

The difference between proposition (d1) and (d2) is in the second conjunct. The second conjunct in proposition (d2) should be interpreted as denying the existence of the entity '¬Nφ' in the past. However, notice that proposition (d2) does not attribute any truth-value to the proposition ¬Nφ, and it is questionable whether it is recognized as a proposition at all. It can simply refer to the combination of written signs that appears between the single quotation marks, which may as well be a meaningless combination of signs, and interpreted as denying that this combination of signs existed in the past. Proposition (d2) thus simply describes a distinction between two objects in the world, a state of affairs in nature and a combination of signs, without implying any connection between the two. In fact, notwithstanding the visual similarity in the signs that are used in the two conjuncts of proposition (d), it is not even clear that there is a conceptual connection between the proposition ¬Nφ, which appears in the first conjunct, and the combination of signs '¬Nφ,' which appears in the second conjunct.

Thus, recognizing the existence or non-existence of propositions as entities in the world does not imply that they are identified as propositions, that is, as entities to which a truth-value can be ascribed. Although this is a necessary condition for explaining the conceptual possibility that is described in proposition (d1), the mere recognition of the existence of the conceptual scheme and its propositions is not a sufficient condition for supporting this possibility.

In what follows, I show that to acknowledge the existence of propositions as entities that can be either true or false, a conceptual scheme must not only include the conceptual apparatus that enables making the distinction between a proposition and what it describes in the world. It must further acknowledge the distinction between the actual truth-value of propositions and the truth-value that is ascribed to them in this conceptual scheme.

To explore the necessary conditions for acknowledging the existence of propositions as propositions, I begin by considering propositions as abstract entities, as a minimal starting point for my investigation. This is because although I have argued that propositions as abstract entities are not sufficient for explaining the need for contingent propositions for describing what exists, it is not clear that it is also necessary to acknowledge propositions as such.

To acknowledge propositions as such, that is, as entities to which truth-values can be ascribed, it is necessary to be able to make the distinction between a true proposition—not merely a proposition to which the truth-value "true" is (perhaps mistakenly) ascribed—and what is described by this proposition. Similarly, it must be possible to distinguish between a false proposition and what is implied by this falsehood about the world. Otherwise, in sentence (b2), that is, P(T'¬(Nφ)')), the combination of signs T'…' could not contribute to the meaning of this sentence, which would therefore be reduced to sentence (b3), that is P(¬(Nφ)). Hence, it would fail to require the use of contingent propositions for describing nature.

To see why, let me map the conceptual space that includes proposition ¬(Nφ) and its possible truth-values on the one hand and what is described by it on the other, as demonstrated in Table 4.7.

Table 4.7 The conceptual space that includes proposition ¬(Nφ)

(4.7.1) ¬(Nφ)	(4.7.3) T'¬(Nφ)'
(4.7.2) ¬(¬(©φ))	(4.7.4) (¬T)('¬(Nφ)')

If propositions were only abstract entities, rather than actual contingent entities in the world, it would have been impossible to make the distinction between propositions (4.7.1) and (4.7.2), which describe the world, on the one hand, and propositions (4.7.3) and (4.7.4), which describe the truth-value of the proposition in language, on the other. Propositions (4.7.1) and (4.7.3), as well as propositions (4.7.2) and (4.7.4), are logically equivalent. Hence, this conceptual space can be mapped with the help of propositions (4.7.1) and (4.7.2) alone. The added reference to propositions and the ascription of a truth-value contribute nothing to the meaning of sentences (4.7.3) and (4.7.4). That is, it is impossible to draw the distinction between propositions and their truth-value and what these propositions describe in the world, as long as propositions are understood to be abstract entities.

Hence, the idea of propositions as contingently existing entities in the world is not only a requirement for explaining the possibility of both affirming and denying the same proposition without a contradiction. It is also required to explain how propositions are distinguished from the things in the world they describe. Once propositions are viewed, not merely as abstract entities, but rather as contingently existing entities, it is possible to begin making the distinction between the proposition and its truth-value and what it describes. For, in this case, propositions (4.7.1) and (4.7.2) do not imply propositions (4.7.3) and (4.7.4) respectively. This is because it is possible that, although propositions (4.7.1) and (4.7.2) are true, propositions (4.7.3) and (4.7.4) are nevertheless false. This is the case if the proposition $\neg(N\varphi)$ is not constituted by any conceptual scheme.

However, this is not sufficient, to distinguish between the truth-value of a proposition and what it describes in the world, to only acknowledge the possibility that the world is as described but the proposition does not exist. Even if this is acknowledged, it is nevertheless possible to view the proposition as a property of these states. That is, it nonetheless leaves open the possibility that the single quotation marks are used to add a distinction to the meaning, that is, the description, of the proposition to which they are attached. To acknowledge the existence of propositions, as entities that are different from what they describe, it is also necessary to recognize the possibility that the propositions exist, but not what it describes.

Furthermore, to acknowledge the existence of propositions as propositions, it is also necessary to acknowledge the ascription of truth-value to these entities. This brings me to the distinction between the actual truth-value of a proposition and the truth-value that is ascribed to a proposition in a conceptual scheme. To identify an entity as a proposition, it is necessary to acknowledge the possibility that this proposition was ascribed with the truth-value "false" while it was actually true, and vice versa, that is, that it was ascribed with the truth-value "true" while it was actually false.

To understand why, recall the claim that to make the distinction between a proposition and what it describes, it is necessary not only to acknowledge the possibility that what a proposition describes exists while the proposition does not, but it is also necessary to acknowledge the possibility that the proposition exists while what it describes does not exist. Notice, however, that this condition cannot be formulated in terms of the propositions of Table 4.7, although it is tempting to interpret this condition as the possibility that proposition (4.7.3) is true while proposition (4.7.1) is false, and (4.7.4) is true while proposition (4.7.2) is false. This is because, in propositions (4.7.3) and (4.7.4), the truth-value that is ascribed to proposition $\neg(N\varphi)$ is its actual truth-value. Hence, propositions (4.7.1) and (4.7.2) follow from propositions (4.7.3) and (4.7.4) respectively. It is therefore impossible to draw this distinction between the proposition and what it describes while ascribing the actual truth-value to this proposition.

Furthermore, it can be shown that acknowledging the existence of propositions as part of the existence of a conceptual scheme, is not a sufficient condition for giving meaning to the ascription of an actual truth-value to a proposition. For, once what exists in the world is described and the existence of a proposition is recognized, the actual truth-value of the proposition can be deduced. Hence, according to the conditions for the meaningfulness of expressions in language, the ascription of an actual truth-value to a proposition is superfluous, that is, meaningless, once what exists in the world is described.

The only way to distinguish between a proposition as a proposition, that is, as an entity to which a truth-value can be meaningfully ascribed, and what it describes, is by referring to the truth-value that is ascribed to this proposition in a conceptual scheme, rather than its actual truth-value. Furthermore, the meaningfulness of ascribing a truth-value to propositions is based, first and foremost, on the truth-value that is ascribed to a proposition in a conceptual scheme, whether this ascription is accurate or not, rather than on the actual truth-value of this proposition. Obviously, the distinction between the actual truth-value and the truth-value that is ascribed to a proposition relies on the premise that this distinction can in fact be drawn, that is, that it is possible to identify an inconsistency between the two. That is, it implies the possibility that the truth-value that is ascribed to a proposition is not its actual truth-value, and that this discrepancy can be identified. This is the possibility that I termed earlier (in Sect. 4.11) "the possibility of mistake."

I might nevertheless be wondering whether the possibility of mistake is indeed necessary for acknowledging the distinction between the truth-value of a proposition and what it describes in the world. Recall, in this context, that while considering (in Sect. 4.11) how to explain the truth of the second conjunct in proposition (d1), that is, $\neg P(T`\neg(N\varphi)')$, I have briefly discussed suggestion (2), according to which the proposition $\neg(N\varphi)$ was either considered false or its truth-value was not decided in its conceptual scheme. The first possibility reflects the possibility of mistake, while the second reflects the possibility of ignorance. The question arises: Is it possible to give meaning to this distinction by acknowledging the possibility of ignorance, that is, by simply acknowledging, for example, that although it was true that $\neg(N\varphi)$, no truth-value was ascribed to the proposition at that time?

The answer is negative. The possibility of ignorance is not sufficient to give meaning to the distinction that is required here. What is needed is to give meaning to the distinction between propositions (4.7.1) and (4.7.2) on the one hand, and propositions (4.7.3) and (4.7.4) on the other hand. Propositions (4.7.3) and (4.7.4) ascribe truth-values to the proposition '$\neg(N\varphi)$,' and hence this distinction cannot be explained by the possibility of ignorance. This does not rule out the possibility that further research will show that the possibility of ignorance is necessary for explaining this distinction. Nevertheless, the possibility of mistake is necessary for making this distinction.

Interestingly, some philosophers, most notably Descartes, have seen fallibility as a shortcoming, and have attempted to rebuild our knowledge in such a way as to eliminate any possibility of mistake. Thus, for example, the attempt was made to base all our knowledge on a present tense first-person account of the content of our consciousness. However, my analysis shows that the concept of a truth-value in general, as well as the idea of a proposition, can only be included in a conceptual scheme due to the possibility of mistake. Hence, the very notion of knowledge is available for this reason—the ascription of knowledge is not meaningful in cases of infallibility. A present tense first-person account of the content of our consciousness (if it is indeed infallible) is therefore not sufficient even to support the concept of truth, let alone ascriptions of knowledge.

The possibility of mistake can be viewed as an implication of the status of propositions in language as contingently existing entities. Hence, the relations between propositions and the things they describe, as a relation between contingent entities, must be likewise contingent. Furthermore, the ascription of truth-value to a proposition, as a contingent occurrence, must likewise be logically independent of what this proposition describes. This implies the possibility of mistake.

In conclusion, to include this distinction, between a proposition and its truth-value on the one hand, and what it describes in the world on the other, a conceptual scheme must acknowledge the possibility of mistake. To acknowledge the possibility of mistake, the conceptual scheme must be able to refer to the truth-value that is ascribed to a proposition, in contrast to its actual truth-value, and acknowledge this discrepancy.

To refer to the truth-value that is ascribed to a proposition, in contrast to its actual truth-value, it is necessary to introduce another conceptual apparatus to the conceptual scheme I am constructing here. So far, I have introduced the signs "T" and "(¬T)" to refer to the truth-value of a proposition. To refer to the truth-value that is ascribed to a proposition, I use the same conceptual apparatus that was used to refer to propositions, that is, single quotation marks. Thus, the distinction between the truth-value that is ascribed to a proposition and what is described by this proposition is demonstrated in proposition (25).

(25) $N\varphi \wedge$ '(¬T)' '$N\varphi$'

According to proposition (25), although $N\varphi$, the proposition '$N\varphi$' is ascribed with the truth-value "false" in this conceptual scheme. However, proposition (25) involves a pragmatic contradiction. The reason for that is that this conceptual scheme must have the conceptual resources to infer that '$N\varphi$' is true, that is, 'T' '$N\varphi$,' from the first conjunct in proposition (25), and hence it makes no sense to assume that it ascribes to this proposition the wrong truth-value.

To solve this difficulty, there is no need to introduce any further conceptual apparatus. This is because the beginning of the present inquiry involves an attempt to solve a similar problem, that is, to solve the contradiction that arises from ascribing different truth-values to the same proposition. The solution for the current difficulty is hence found again in the temporal indexical ability to contain the propositions in question, thus avoiding the contradiction, as exemplified in proposition (26).

(26) $P(N\varphi) \wedge P(`(\neg T)' `N\varphi')$

According to proposition (26), although the proposition "now φ" was true in the past, in this conceptual scheme it was wrongly ascribed with the truth-value "false." Proposition (26) therefore demonstrates a conceptual scheme's acknowledgment of the distinction between what exists and the proposition it uses to describe what exists, and acknowledgment of the possibility that the truth-value it ascribed (in the past) to this proposition might be wrong. Notice that although this proposition relates to the past, due to its use of the proposition "now φ" it implies that the present ought to be independent of its description by this conceptual scheme, and susceptible to the possibility of mistake. Thus, although it is impossible in the confines of this conceptual scheme to identify any mistake in attributing truth-values to propositions about the present, this conceptual scheme can acknowledge the possibility that this is the case, and that this would be discovered in the future.

So far, I have discussed a proposition about the present, that is, Nφ. However, the same considerations equally apply to propositions about the past, that is, Pφ. That is, to identify a proposition about the past as a proposition, a distinction must be made between the truth-value that is ascribed to this proposition and what is described by this proposition, as exemplified in proposition (27), which demonstrates that the past should be independent of its description by the conceptual scheme.

(27) $P(P\varphi) \wedge \neg P(`(\neg T)' `P\varphi')$

Again, the conceptual scheme must be able to acknowledge the possibility of mistake regarding the truth-value of propositions about the past, to identify them as propositions. That is, the conceptual scheme must recognize the possibility that the past is different from the way it is described by it. Again, this is only possible without a contradiction when applied to the past.

Lastly, the considerations that I have presented so far also imply the possibility of ignorance. This is because the conceptual scheme in question must not only include the possibility of mistake, but further acknowledge this possibility. However, if the conceptual scheme acknowledges that it might be mistaken about the correct truth-value of its propositions, it is reasonable to assume that there are circumstances in which the conceptual scheme would acknowledge that there are things that are not known with certainty, and hence suspend judgment on the truth-value of some of its propositions. This is obviously the possibility of ignorance.

Admittedly, as long as I have not given an account of objectivity, it is unclear how the possibilities of mistake and ignorance can be acknowledged, and how such concerns should be dealt with and resolved. A solution to these queries requires an account of objectivity, a task that is undertaken in Chap. 5.

In conclusion, acknowledging the propositions in language as propositions requires a conceptual scheme to acknowledge the distinction between its descriptions of nature and what is actually the case. In other words, this distinction relies on an objective conception of nature.

4.13 The Constitution of Time

In Table 4.6 I described the conceptual structure that is made available by the context operators. However, I must remind you that this table does not constitute a temporal series of occurrences from the near past to the far past. This is because lower-order propositions, such as proposition (4.6.2), do not describe the near future, but are general propositions that claim that somewhere in the past φ was realized. Hence, proposition (4.6.2) follows from any higher-order proposition.

I have identified the context operators with the temporal indexical expressions in ordinary language. However, I have stressed that, by this identification, it was not my intention to import into the discussion the complex conceptual structure that is associated with the distinctions between past, present, and future in our conceptual scheme. Every conceptual apparatus that is introduced into the current discussion must be justified based on the conceptual structure that has been established so far. Two such aspects of our conception of time remain to be introduced now, to justify my identification of the context operators with the temporal indexical expressions. These are the structure of time and the notion of the future.

It is now appropriate to show how the conceptual structure that was introduced so far enables constituting time as a linear dimension of occurrences. To begin with, recall that this conceptual apparatus enables the conceptual scheme to document the change in the truth-value of propositions, including propositions about the present. Hence, Table 4.6 can be formulated in a way that records the change in what occurs in the present, as demonstrated in Table 4.8.

I shall explain in Chap. 5 why this structure must be founded on propositions that describe the present. What interests me now is the following question. What justifies the addition of the higher-order indexical expressions to the basic proposition $N\varphi$?

If the proposition $N\varphi$ was once affirmed in a conceptual scheme, it is recorded by proposition 4.8.2, that is, $P(N\varphi)$. Similarly, the fact that at some point in time the proposition $P(N\varphi)$ was true, is correctly recorded in later moments as $P(P(N\varphi))$. However, this change in time cannot be documented by recording the passage of time, that is, the change in the distinctions of past, present, and future. For, as I have argued, context operators are not ordinary distinctions. They create a context for the

Table 4.8 The change in what occurs in the present

4.8.1	$N\varphi$
4.8.2	$P\varphi \equiv P(N\varphi)$
4.8.3	$P(P\varphi) \equiv P(P(N\varphi))$
4.8.4	$P(P(P\varphi)) \equiv P(P(P(N\varphi)))$
	…

determination of other distinctions. This explains why the temporal indexical distinctions are not identified directly; indeed neither are any temporal distinctions that will be shown in what follows to be constituted based on the temporal indexical expressions. They reflect the fact that nature is described from a point of view that is located in nature, rather than contributing to the content of what is described from this point of view.

It follows from this that it is impossible to record a change in time unless there is a change in the content of the propositions that describe nature. It might be thought, therefore, that it is only possible to record a change in time by documenting a change in the properties that are included in φ. However, recall that I have argued that, to explain the ability to ascribe conflicting truth-values to a proposition without a contradiction, a conceptual scheme must be able to identify its propositions and their truth-values. Hence, the change that is needed to describe the passage of time is also satisfied by the change in the truth-values of the propositions of the conceptual scheme. Thus, the proposition Nφ need not change its truth-value in time for a conceptual scheme to record the passage of time. Recording its contents suffices for recording occurrences in the form of a linear series, as described in Table 4.9.

Notice that the change that is documented in Table 4.9 does not involve any change in the property φ, which remains realized throughout. What is documented is the change in truth-values of propositions in the conceptual scheme. Further, notice that the series of propositions from 4.9.2 onwards form a series in time. However, the series of occurrences in time that is documented by the propositions are included in each proposition in this series, in which the descriptions in the inner parentheses describe earlier occurrences. Thus, the series of occurrences in time that is documented by this conceptual scheme gets longer with each proposition that is added, thereby describing what is experienced as a change in the properties of present and past, which constitutes the "flow of time," and forms a linear series of occurrences in time.

I point out that this conceptual structure, because of the possibility of time in which the conceptual scheme did not exists (which was found necessary for explaining the contingency of propositions in language), allows for gaps in the series of occurrences. That is, it is a subjective series of occurrences, as they are recorded from a limited point of view of nature. I have argued, however, that the conceptual structure that is constructed in this chapter requires an objective conception of nature.

Table 4.9 The change in truth-values of propositions in the conceptual scheme

4.9.1	Nφ
4.9.2	Nφ ∧ P(T'Nφ')
4.9.3	Nφ ∧ P(T'(Nφ ∧ P(T'Nφ'))')
4.9.4	Nφ ∧ P(T'(Nφ ∧ P(T'(Nφ ∧ P(T'Nφ'))')')')
	…

Furthermore, although I have described this series as a temporal one, I must emphasize that I am not introducing the complex conceptual structure associated with the temporal dimension in our conceptual scheme. In this respect, I acknowledge that this is not a "full-blooded" temporal series, as the structure uncovered thus far pales in comparison to the complexity of the structure underlying temporal distinctions in our actual conceptual scheme. This deeper analysis, incorporating the objective temporal relations of simultaneity and precedence, and the formation of an objective temporal series, will be taken up in Chap. 6.

Even though, in our conceptual scheme, propositions about the future must be inferred, and are not directly observed, it is possible to use the conceptual structure that was developed so far to introduce the temporal indexical expression "future," although only in a relative use. Yet to be explained is how this conceptual structure can enable referring to actual future occurrences, that is, to occurrences that are later than the actual present. At this stage, I only show how it is possible to use the notion of the future in relation to past occurrences. This suffices to give meaning to the idea of the future.

To introduce the notion of the future into the conceptual structure that was developed so far, notice that it is simply the reverse of the notion of the past. That is, an occurrence φ is in the future in relation to a point in time T, if and only if, when φ is present, T is in the past. This resolves the problem of explaining how language includes the three temporal indexical expressions of past, present (now), and future, while my analysis only required two context operators to explain the need for contingent propositions for describing what exists—the third indexical expression is implied by the previous more basic two.

Finally, propositions that include the distinctions between past, present, and future are found to be the fundamental propositions in language. Hence, it is necessary to explain how natural language can include propositions that are not sensitive to the temporal context of their use, and hence have a fixed truth-value. Although most propositions in language are sensitive to the temporal context of their utterance, due to tense, there is an atemporal use of the present tense, for example in the case of mathematical propositions (Frege, 1984, p. 358). Such uses can be introduced into the language, based on the distinctions between past, present, and future, by the stipulation that a proposition of the type "*a is* S," in which the italicized "*is*" represents the atemporal use, is interpreted as "*a* was S, or *a* is now S, or *a* will be S."

The constitution of the dimension of time, as a linear series of occurrences, concludes this chapter. Implications of this conceptual structure, which explains and enables the use of contingent propositions for describing nature, will be analyzed and developed in the following chapters.

References

Brogaard, B. (2012). In B. Brogaard (Ed.), *Transient truths: An essay in the metaphysics of propositions*. Oxford University Press.

Cappelen, H., & Dever, J. (2013). *The inessential indexical: On the philosophical insignificance of perspective and the first person*. OUP Oxford.

Dolev, Y. (2000). Space and time: Some (dis)analogies. *Iyyun: The Jerusalem Philosophical Quarterly, 49*, 61–72.

Frege, G. (1948). Sense and reference. *The Philosophical Review, 57*(3), 209–230.

Frege, G. (1984). Logical investigations: Thoughts (P. Geach & R. H. Stoothoff, Trans.). In B. F. McGuinness (Ed.), *Collected papers on mathematics, logic and philosophy* (pp. 351–372). Basil Blackwell.

Gale, R. M. (1969). "Here" and "now". *The Monist, 53*(3), 396–409.

Hardin, C. L. (1984). Thank goodness it's over there! *Philosophy, 59*(227), 121–125.

Kaplan, D. (1989). Demonstratives. In J. Almog & J. Perry (Eds.), *Themes from Kaplan* (pp. 482–563). Oxford University Press.

Mellor, H. D. (1998). *Real time II*. Routledge.

Perry, J. (1977). Frege on demonstratives. *Philosophical Review, 86*(4), 474–497.

Perry, J. (1979). The problem of the essential indexical. *Noûs, 13*(1), 3–21.

Russell, B. (1956). The philosophy of logical atomism. In *Logic and knowledge*. George Allen & Unwin.

Taylor, R. (1955). Spatial and temporal analogies and the concept of identity. *Journal of Philosophy, 52*(22), 599–612.

Wittgenstein, L. (1963). *Tractatus logico-philosophicus* (D. P. Pears & B. F. McGuinness, Trans.). Routledge & Kegan Paul.

Yehezkel, G. (2014). *The conceptual structure of reality*. Springer.

Chapter 5
The Structure of the Laws of Nature

Abstract In this chapter I explain how a conceptual scheme can formulate an objective description of nature from a limited point of view in time. I show that this notion of objectivity, which is implied here, ultimately relies on the existence of laws of nature, and further uncovers their necessary structure. Nature is found to be necessarily subject to laws of nature, which connect occurrences at different times, such that occurrences at one time are both necessary and sufficient conditions for occurrences at other times. The structure of the laws of nature is further developed and explained, and it is shown that all occurrences in nature must be reciprocally connected by these laws.

Keywords Contingency · Determinism · Laws of nature · Objectivity

5.1 Introduction

The conclusion of Chap. 4 is that the possibility of ascribing different truth-values to the same proposition, which also explains the need for contingent propositions for describing nature, requires the distinction between propositions and what they describe. This in turn implies that a conceptual scheme must incorporate a notion of objectivity. This includes acknowledging the limited existence of this conceptual scheme, and hence of its description of nature, as well as the possibility of mistake, that is, of the possibility of ascriptions of wrong truth-values to propositions in this conceptual scheme. This applies to any proposition, whether it is about the past, the present, or the future.

I should emphasize again that the identification of the context operators with temporal indexical expressions does not imply endorsing the complex and rich conceptual structure that is associated in our conceptual scheme with time. So far, the conceptual structure that has been uncovered is mainly encapsulated in Table 4.9 and the relations between propositions (a2), (b2), and (b3).

In this chapter, I discuss some of the necessary conditions for supporting these conceptual distinctions. I focus mainly on the idea of objectivity, and offer an

G. Yehezkel, *Conceptual Principles of Natural Philosophy*, Synthese Library 495, https://doi.org/10.1007/978-3-031-82108-0_5

account of objectivity in terms of laws of nature. By exposing the role of the laws of nature in supporting an objective conception of nature, I am able to determine their necessary structure. Although many of the conclusions of this analysis may seem trivial to anyone familiar with the development of physics since the beginning of modern science, this investigation establishes many prevalent views about the structure of the laws of nature, that were thus far based on intuition alone, as necessary features of nature.

I use the term "laws of nature" to refer to a specific type of contingent regularities that obtain in nature. Although they are contingent regularities, the laws of nature are distinguished from mere accidental regularities (Goodman, 1983, p. 73). The problem of distinguishing between accidental regularities and laws of nature is challenging (see, for example, (Dretske, 1977; Jackson & Pargetter, 1980; Lange, 2000, pp. 111–112)). However, according to my analysis, the distinction between laws of nature and accidental regularities is secured by the fact that the existence and structure of the laws of nature are necessary features of nature, and implied by the conceptual structure of time, as was revealed in Chap. 4. My analysis shows that all occurrences in nature must be reciprocally connected by the laws of nature, which tie together occurrences at different times, such that occurrences at one time are both necessary and sufficient conditions for the occurrences at another time.

Finally, I should warn against misunderstanding the account of the laws of nature that I am developing. The subject of the laws of nature has been discussed extensively in philosophical literature, especially in the context of the philosophy of science. As mentioned earlier, it is not my intention to analyze the laws of nature as they appear in scientific theories—an analysis which is the subject matter of a philosopher of science, rather than that of a natural philosopher. Hence, I shall not attempt to enter the debate on the many disputes in this field. These include, inter alia, antirealist views about the laws of nature (Giere, 1999; Mumford, 2004; Van Fraassen, 1989; Ward, 2002) and two main competing views about what constitutes a law of nature, that is: the "systems approach," which explains the laws of nature in terms of deductive system (Earman, 1984; Lewis, 1973, 1983, 1986, 1994; Loewer, 1996); and the "universals approach," which relies on universals to explain what constitutes a law of nature (Armstrong, 1978, 1983, 1991, 1993; Tooley, 1977, 1987).

According to my conception of the laws of nature, the laws of nature constitute the objective realm, rather than describing some of its features. Hence, for example, my view is incompatible with "Humean Supervenience," according to which everything, including the laws of nature, supervene on particular facts (Lewis, 1986, p. ix). As will become apparent in what follows, the view that emerges from my analysis is a realist view and in some respects is akin to the "universals approach." However, notwithstanding the superficial similarity, it is important not to confuse the view I am advancing here with any of the common conceptions of the laws of nature. The view that will emerge in this chapter is formulated within the confines of a unique conception of nature (which will be further developed in the following chapters). It is based, for example, on a notion of conceptual necessity, and a unique combination of necessity and contingency (which will be further discussed in Sect.

5.15), which separates it from all the conventional views of the laws of nature. Thus, for example, although according to this view there is a sense in which the laws of nature are necessary truths, it should not be confused with traditional necessitarian views (see, for example, Bird, 2005; Fales, 1990; Shoemaker, 1980, 1998; Swoyer, 1982).

5.2 The Distinction Between the Subjective and the Objective Realms

The analysis of Chap. 4 has highlighted the importance of the notion of objectivity, as reflected in the distinction between propositions and what they describe. This distinction was shown to imply the possibility of mistake (and hence also of ignorance); that is, the possibility of disparity between what actually exists and what is described by the conceptual scheme.

The distinction between propositions and what they describe must apply to all propositions, that is, propositions about the past, present, and future. This is because, as I have argued before, it is impossible to identify them as propositions if they are not distinguished from what they describe, and this in turn requires acknowledging the possibility of mistake.

As was explained in Chap. 4, propositions that include the distinctions between past, present, and future describe nature from a point of view that is located in the present. For example, the proposition "It was raining in the past" is equivalent to "Now it is true that 'It was raining in the past'." However, it is impossible to acknowledge a mistake regarding propositions in the present, as demonstrated in proposition (28).

(28) $N\varphi \wedge$ '$\neg T$' '$N\varphi$'

Proposition (28) can be compared to a subject who claims both that "it is raining now" and that "I think that the proposition 'it is raining now' is not true." If in a conceptual scheme a proposition is affirmed, it makes no sense that it is also considered in this conceptual scheme to be "not-true." The solution for this problem was found in the very same conceptual structure that explained the need for contingent propositions for describing what exists, as described in proposition (29).

(29) $P(N\varphi) \wedge P($'$\neg T$' '$N\varphi$'$)$

Proposition (29) describes the present acknowledgment that a proposition about the (then) present was considered in the past to be false, but was (found later to be) actually true. Similar propositions can be formulated about past and future propositions, as is demonstrated in propositions (30) and (31).

(30) $P(P\varphi) \wedge P($'$\neg T$' '$P\varphi$'$)$
(31) $P(F\varphi) \wedge P($'$\neg F$' '$F\varphi$'$)$

Thus, propositions (29) to (31) explain how propositions about the past, present, and future can be distinguished from what they describe, and thereby establish the idea that the past, the present, and the future are objective.

An implication of this is the existence of a subjective realm. As has been shown, it is impossible to distinguish between an ascribed truth-value and what actually is the case regarding propositions that describe nature (only) from the present. Notice however that this status is limited to *the state of the conceptual scheme in the present*. That is, for any type of proposition that in the present is ascribed a truth-value in a conceptual scheme, it is impossible to acknowledge (in the present) that this ascription is false.

This conclusion can be summed up thus: everything in the past and the future is objective, that is, susceptible to the possibility of mistake, including the state of the conceptual scheme itself. That is, it is possible to formulate false descriptions of anything in the past and the future, including descriptions of what is held as true or false in this conceptual scheme. Thus, for example, it is possible that the proposition P('¬T' 'Fφ'), which describes the state of the conceptual scheme in the past, is held now as true, although it is actually false, and in fact in this conceptual scheme the proposition Fφ was never held as not-true. The complete objectivity of the past and the future is not surprising given the fact that the conceptual scheme was found to exist as part of nature, and hence its states in the past and the future are susceptible to the possibility of mistake.

In the present, on the other hand, there is an objective realm, which is susceptible to the possibility of mistake, as demonstrated in proposition (29). However, there is also a subjective realm, which includes only the states of the conceptual scheme itself. This realm is subjective in the sense that it is not susceptible to the possibility of mistake, on pain of self-contradiction—as demonstrated in proposition (28). The coherence that is required of a conceptual scheme does not allow opening the gap that is needed for the possibility of mistake.

Any conceptual scheme must hence include a class of propositions that are not susceptible to doubt, and in fact must be infallible, to maintain the consistency of the conceptual scheme. These are the propositions that describe the state of the conceptual scheme in the present, that is, its propositions and their ascribed truth-value. Moreover, this unique status should be acknowledged by the conceptual scheme itself. For, as I have argued earlier, it is necessary for a conceptual scheme to acknowledge its existence as such, as well as its propositions and their ascribed truth-values, to explain the need for contingent propositions for describing what exists. Hence, regarding any proposition *p*, that is ascribed the truth-value "true" in a conceptual scheme, this conceptual scheme must be able to acknowledge that N('T' '*p*').

A few explanatory remarks are in order. To begin with, I do not merely argue that in this conceptual scheme the proposition "T*p*," that is, "it is true that *p*," is held true. To clarify, this is indeed the case. However, this proposition is susceptible to the possibility of mistake, because the proposition *p* itself might be susceptible to the possibility of mistake. Furthermore, since the conceptual scheme can acknowledge its own mistakes in the past, it must acknowledge the possibility that the truth-value it now ascribes to the proposition *p* may be wrong. My claim is that this conceptual scheme must acknowledge *its ascription of a truth-value* to the proposition *p*, whether it is wrong or not, and that this acknowledgment is not susceptible to mistake, given the coherence of this conceptual scheme. That is, the proposition N('T' '*p*') should be interpreted as claiming that the proposition *p* is currently *accepted as true* in this conceptual scheme. This reflects the fact that this conceptual scheme describes nature from the present, and the fact that this conceptual scheme acknowledges its existence as a conceptual scheme.

Furthermore, I should emphasize that this acknowledgment does not require any unique way of gaining descriptions of nature—a kind of perception of its states, which is sometimes termed "reflection" or "introspection." It relies on a conceptual structure that was found necessary for explaining contingency and hence meaningfulness in general. This demand is reflected in our conceptual scheme by Moore's paradox, that is, in the absurdity of the claim "*p*, but I believe that not-*p*," which reflects our awareness of our use of propositions to describe what exists.

5.3 Inferential and Non-inferential Descriptions

Although the previous section has shown that any conceptual scheme must include infallible propositions, they can only describe the content of the conceptual scheme itself. Hence, although some philosophers have envisioned that all our knowledge, including our knowledge of nature, will be constructed on an infallible basis, and will therefore be infallible throughout, this project is not feasible. This is not only because this offers a very limited basis, which includes only the conceptual scheme itself and the truth-values that are ascribed to the propositions that are formulated in its confines, rather than what these propositions are supposed to describe, that is, nature. This is also because contingent propositions, and hence meaning in general, were shown to be founded on propositions that are susceptible to the possibility of mistake.

This leads me to the question, how can is this type of description of nature, that is, the type that is susceptible to the possibility of mistake, be acquired? In our conceptual scheme, it is possible to make the distinction between descriptions of nature that are deduced based on other propositions that are held as true, and those that are non-inferential, and conceived as a straightforward description of nature, that is, by what in our conceptual scheme is termed "experience." There are obviously more distinctions and clarifications to be made in our actual complex conceptual scheme. Thus, for example, some propositions are ascribed a truth-value based on sense perception, and hence are susceptible to the possibility of mistake. There are propositions, on the other hand, that describe our psychological states, and those were traditionally thought, at least by some philosophers, to be infallible—analogous to the conclusion of the previous section. However, I shall not delve into our complex conceptual scheme, and simply rely on the basic distinction between inferential and non-inferential descriptions of what exists. It is nevertheless important to clarify that, although in what follows I do not discuss or rely on the notion of knowledge, due to its complex relations with ideas that have yet to be introduced into this investigation, the idea of a non-inferential description of nature ought to be understood as a justified ascription of truth-value to a proposition. That is, a non-inferential description cannot be an arbitrary adoption of a proposition as a true description of nature, for this would undermine the ideas of truth and (descriptive) meaning. Hence, whenever I discuss the idea of a non-inferential description of nature, it should be understood that this requires a suitable connection (in a yet undefined sense) between the description and what it describes.

For there to be contingent propositions that are held as true, there must be contingent non-inferential propositions that are held as true. This is because inferential propositions must follow from other propositions, and there must be basic propositions on which any structure of inferences is supported. Admittedly, it is possible that knowledge has no fundamental level, and is comprised of propositions that cohere with one another. However, in this context, these also should be considered to be non-inferential. Furthermore, contingent propositions cannot be deduced from necessary propositions alone. Hence, the basic descriptions of nature must be non-inferential contingent propositions.

Non-inferential contingent propositions that are held as true are therefore epistemologically basic, which does not imply that they are infallible. For, as I have argued, the basic propositions in language, that describe what exists with the help of the distinctions between past, present, and future, are in fact fallible, because they are susceptible to the possibility of mistake.

In the previous chapter, I began developing a conceptual scheme that includes contingent propositions. This conceptual scheme must include, as I have just argued, non-inferential contingent propositions that are held as true. In our actual conceptual scheme, however, there are also inferential propositions that are held as true, such as general propositions that allegedly describe laws of nature, and propositions that describe particular facts, which are derived from general propositions together with other propositions that describe particular facts. The question that arises is whether a conceptual scheme, which has the conceptual resources to formulate contingent propositions, must also have the conceptual resources to support the possibility of contingent inferential propositions.

The analysis of the previous chapter has shown that propositions that include the temporal indexical expression of past and present are the basic propositions in language, and hence also the basic descriptions of nature. Hence, at least some propositions of this type must be non-inferential, and hence accepted without being implied by other propositions. In other words, since these are the most basic descriptions of what exists, at least some of them must be used in a straightforward description of what exists, rather than be implied by other descriptions.

In Sect. 4.10 I identified the context operators © and (¬©), which were constructed as an abstract conceptual apparatus that enables the same proposition to be both true and false without a contradiction, with the temporal indexical expressions "now," or "present," and "past." The justification for this identification was twofold. The context operator © was identified with "present" due to its unique function as the default context operator, which fits the function of the temporal expression "present." The context operator (¬©) was identified with "past" because (¬©) was identified as a basic distinction in language. I have presented a brief justification for this identification in Sect. 4.10, but at this stage of the inquiry, I can offer further justification for this identification.

According to the conception of language as a system of coordinates, to give meaning to a distinction, a conceptual dimension must include at least two conceptual possibilities. This was the consideration that led me to the introduction of the context operators (¬©), following the introduction of the context operator ©. Furthermore, this stipulation was found necessary to explain the possibility of ascribing different truth-values to the same proposition. Because the two context operators depend on one another, they are equally basic, and it is impossible to define one based on the other. (It is possible, however, to define other context operators based on the two basic operators—as proved by the introduction of the temporal indexical expression "future.")

The previous discussion has shown that some propositions must be accepted non-inferentially. These are the basic descriptions of nature, which form the basis for the inferential descriptions of nature (if such descriptions are possible). The fact that in our conceptual scheme propositions that describe nature with the help of the indexical expression "future" are inferential implies that the temporal indexical expression "future" cannot be identified as one of the two basic context operators. On the other hand, although it is possible to derive conclusions about the past, there are in our conceptual scheme propositions about the past that are not the conclusions of inferences, that is, memories. This indicates that the indexical expression "past" should be identified as one of the two basic context operators that were constructed in the previous chapter.

My conclusion is that in every possible conceptual scheme, some propositions that describe nature with the temporal indexical expressions of "present" and "future" must be accepted as true non-inferentially. That is, they must be conceived as directly describing nature, rather than be implied by other propositions. I should again emphasize that this does not imply that all the propositions that describe nature with these distinctions represent a direct description of nature. I only argue that some of them must be accepted as true non-inferentially, for these are the basic propositions in language, on which all other contingent propositions in language rely, both for their meaning and for describing nature (which are tied together, as I have argued in Chap. 2).

5.4 The Problem of Describing the Past

I first introduced the idea of a context operator ©, which was later identified with the temporal indexical expression "present," to explain how a proposition of the structure ©φ can be both true and false. The idea was that the distinctions that are included in φ depend on the context operator. That is, although the distinctions that are included in φ apply to what exists, it is the context that determines how. Nevertheless, the distinctions that are included in φ determine the truth-value of the proposition ©φ.

The context operator © hence contributes to the determination of the truth-value of propositions through its impact on the distinctions that are included in φ. It does not describe what in our conceptual scheme is considered to be directly accessible for description. It is a context for the applicability of distinctions that describe what is directly accessible. Hence, even though it is the basic conceptual structure that enables giving meaning to expressions in language, and thus represents a fundamental and necessary feature of nature, it can be described as a second-order distinction. In terms of our conceptual scheme, it can be said that it does not describe the content of our experience—not a surprising conclusion given that contingent propositions are supposed to describe what contingently exists, rather than the necessary structure of nature.

In order to avoid any confusion, I should stress that the investigation I am embarking on here should not be mistaken as a phenomenological investigation into the nature of the experience, or perception, of time (see, for example, (Poppel, 1978)). Rather, the analysis I am undertaking here is conceptual, and intended to explore the implications of the conceptual structure that was developed in Chap. 4. Although I believe the conclusions of this analysis could shed light on our experience of time, the reverse does not hold. Conceptual analysis should not be conducted based on empirical data.

The temporal indexical distinctions between past, present, and future are hence not what in our conceptual scheme is considered to be directly observable. This is not because they are conceptual constructs, and hence are unobservable, for everything that exists can be said to be a conceptual construct. What it is correct to say is that they are the most basic conceptual construct, that is, a necessary structure of our experience, and hence unobservable. Our experience can only include what is contingent (Yehezkel, 2013).

One may insist that it is possible to observe the "present"—it is just what is given to us directly. This claim, however, is false. One way to acknowledge the fact that the distinctions that are made by the temporal indexical expression are unobservable is to ask how can the "present" be distinguished from the "past." The idea of a distinction implies that there are multiple things in which distinctions can be made. If "being in the past" is unobservable, so is "being in the present." The difficulty of observing the past is that, since what exists is described from the default point of view, which is identified with the present, all the distinctions that can be made are about what exists now.

In this context, it is important to remember the distinction between inferential and non-inferential descriptions. By claiming that the distinctions between past, present, and future are unobservable, I do not deny that these distinctions apply to non-inferential descriptions. Recall that I have argued in Sect. 5.3 that the basic distinction between past and present must be used in non-inferential descriptions. I do argue that, had David Hume analyzed these distinctions with his critical empiricist principles, he would have undoubtedly concluded that they are the product of our imagination. However, Kant has taught us that our concepts can be of complex conceptual structure, rather than simply describe our intuitions, and some of these structures are necessary features of our experience. Hence, paradoxically, for Kant time was also unobservable (Kant, 1929, p. B219).

The question hence arises, how can the past be described? The problem is that my analysis of (descriptive) meaning began with a naïve assumption that describing the world raises no special difficulty, or, as it is sometimes described, made from "the point of view of god." However, the idea of objectivity, that is, that what exists may differ from what is described from the point of view of a conceptual scheme, coupled with the idea that nature is described from the point of view that is located in the present, implies restrictions on the possibility of describing what exists. First and foremost, it seems to imply that everything available for description must be in the present, and the past should be altogether inaccessible.

The problem is hence to explain how it is possible to both describe nature exclusively from a point of view that is located in the present, and to include in our conceptual scheme non-inferential descriptions of the past, that is, descriptions from a point of view that is supposed to be available for a point of view that is located in the past.

The last point can be demonstrated using the conceptual structure that was developed in Chap. 4. Recall that a proposition about the past, that is, $P\varphi$, is logically equivalent to the proposition $P(N\varphi)$, that is, it is equivalent to a proposition that describes the past from a point of view that is located in the past, in relation to which these occurrences are in the present. Hence, if the distinctions that are made by φ are examined now, the resulting description should be of the present, rather than the past. However, it seems that they can only be examined in relation to the present, because the default context operator implies that $P(N\varphi)$ is logically equivalent to $N(P(N\varphi))$.

It might be thought that there is an easy solution to this difficulty. To begin with, the logical equivalence between $P\varphi$ and $P(N\varphi)$ suggests that the origin of the description of the past is in past descriptions of the present. Furthermore, recall that this conceptual structure was shown to imply that the conceptual scheme in question can describe its propositions and their truth-value. In particular, my analysis showed that a conceptual scheme must be able to describe the actual change in the truth-value of the proposition $N\varphi$, a change that reflects the fact that the point of view from which nature is described changes in time. Hence, given the fact that the proposition $P\varphi$ follows from proposition $P(T`N\varphi')$, it seems that the description of the past relies on past descriptions of the present.

In summary, the suggestion for solving the problem of describing the past is that non-inferential descriptions of the past rely on the fact that the point of view from which nature is described, that is, the present, changes in time, coupled with the fact that the conceptual scheme can document this change. All that is required is to record the change in the truth-value of propositions that describe the present, and descriptions of the past will follow.

Notice that this suggestion contradicts the previous claim, that at least some propositions about the past must be non-inferential, due to their fundamental status in language. I shall nevertheless explore this suggestion, to offer a better understanding of the problem I am facing now, and why the distinction between past and present is fundamental.

The problem with the suggestion that propositions about the past follow from recording the change in the truth-value of propositions about the present, is that the notion of the past is supposed to arise from recording the change in the truth-value of propositions that describe the present. However, the notion of the past was introduced specifically to avoid the contradiction that arises from the attempt to ascribe different truth-values to the same proposition. That is, it is impossible to explain how the same proposition can be ascribed with conflicting truth-values without a contradiction, if the idea of the past is not presumed, and included as part of this description.

A simplistic account maintains that the proposition Nφ is true in one moment, and false in the next, and that this is sufficient for acknowledging change. The shortcomings of this suggestion can be demonstrated by the following consideration. Suppose that, in a single conceptual scheme, the proposition Nφ was ascribed at one time with the truth-value "true," and at another time it is ascribed with the truth-value "false." Ascribing conflicting truth-values to a proposition is a straightforward contradiction. This is why Kant argued that time must be a priori, and cannot be abstracted from observed changes (Kant, 1929, p. B48).

The attempt may be made to resolve this contradiction by stipulating that the first ascription of the two conflicting truth-values represents what happened in the past. However, for this suggestion to be applicable, the notion of "first" must have a temporal implication, that is, it must mean "former." That is, it is based on the idea that it is possible to determine which ascription precedes and which follows, that is, which describes the past and which describes the present. Rather than showing how the notion of the past arises from descriptions of the present, this suggestion hence begs the question.

The last point is important because it shows that, even ignoring the problem of contradiction, it is not clear how the distinction of "past" is supposed to be introduced by the ascription of conflicting truth-values to the proposition Nφ. It certainly does not logically follow from these ascriptions. Hence, as I have argued earlier, the description of the past must be, in principle, possible non-inferentially.

In conclusion, it seems that the past is not available for description from the point of view of the present. However, it does not seem to be available by inference from propositions about the present. Hence the difficulty, how can the past be described?

5.5 Non-inferential Description of the Past

I have argued that at least some of the basic descriptions in language, which include the temporal indexical expressions "present" and "past," must be non-inferential. This conclusion was reinforced by the conclusion of the analysis of Sect. 5.4, about the problem of describing the past. The question is, how can there be non-inferential descriptions of the past if nature is described from a point of view that is located in the present? However, before answering this question, I wish to stress that the question that I am asking here is not psychological (see, for example, (Friedman, 1990)). Rather, I am asking a conceptual question, about the conceptual structure that is required for supporting the notion of the past. Hence, for example, relying on "a feeling of 'pastness'," will not be accepted as a valid explanation (Russell, 1921, p. 162).

To solve the problem, it is necessary to return to the idea of the flow of time, that is, the change in the temporal point of view of nature. The conceptual scheme that I am developing here was shown to exist in time, and describes nature from a changing point of view in time (although all these positions are referred to as "now.") This change is demonstrated, for example, in proposition (32).

(32) $N\varphi \wedge P((\neg T)\text{'}N\varphi\text{'} \wedge \text{'}(\neg T)\text{' '}N\varphi\text{'})$

According to proposition (32), while proposition $N\varphi$ is true now, in the past it was false, and it was recorded as such in the conceptual scheme. This implies that this conceptual scheme was once located in what now is the past. Its location in the past allowed this conceptual scheme to describe non-inferentially what was then the present, thus explaining how describing the past is possible.

In this context, it is interesting to see again how much the ability to identify propositions is important for explaining the use of contingent propositions. I have argued before that the ability to identify propositions and their ascribed truth-value is a necessary condition for making the distinction between the propositions and what they describe, and for explaining the possibility of ascribing different truth-values to the same proposition. The current analysis shows another way in which the identification of propositions, in contrast to what they describe, is important. This is because a non-inferential description of the past is found to be available only from the point of view that is located in the past.

So far, I have followed the suggestion that was considered and rejected in the previous section. According to this suggestion, the fact that the proposition $N\varphi$ was in the past recorded as false explains how now the proposition $P((\neg T)\text{'}N\varphi\text{'})$, and hence also the proposition $P\varphi$, are accepted as true. However, while the suggestion that I rejected in the previous section maintained that this connection is logical, it is not. For the proposition $P((\neg T)\text{'}N\varphi\text{'})$ does not follow from the proposition $\neg N\varphi$. This connection hence needs to be explained.

The connection between the ascription of the truth-value "false" to $N\varphi$ in the past to the acceptance of $P((\neg T)\text{'}N\varphi\text{'})$ as true now cannot be logical, and hence it must be a contingent connection. That is, it is not a connection between the content of the two propositions, but a contingent connection between two occurrences, which happen to occur in a single conceptual scheme. These are the ascription of a truth-value to a proposition in the past and the present adoption of another proposition. Hence, a necessary condition for the ability to describe the past non-inferentially is the existence of contingent connections that obtain between occurrences at different times in the conceptual scheme.

Notice that there is an asymmetry between the two occurrences. For according to this explanation, the previous occurrence is supposed to explain the later occurrence. That is, the acceptance of a description of the past in the present is explained by the acceptance of the previous description, and this relates not only to the existence of the two occurrences, but also to the content of the propositions which are involved in this connection. The existence of the previous occurrence, on the other hand, does not depend on the existence of the later occurrence. The existence of the previous occurrence is hence a necessary condition for the existence of the later occurrence, but not vice versa.

To explain the need for contingent propositions for describing nature, it must be possible to formulate non-inferential descriptions of the past, and this, in turn, requires the existence of contingent connections between occurrences at different times, such that the existence of previous occurrences explain, and hence are necessary conditions for, the existence of later occurrences. Hence, at least some occurrences in nature, that is, occurrences in the conceptual scheme from which nature is described, must be connected by contingent connections. Notice that I do not describe this connection as a causal connection. This is because the idea of a causal connection, as it is used in ordinary language, is complex, and involves pragmatic aspects that have no place or justification in the confines of the conceptual scheme I have developed so far.

A few clarifications are required at this stage. To begin with, I do not argue that a contingent connection is required as a necessary condition for justifying claims for *knowledge* of the past. That is, it is not meant to justify claims about the past. As already mentioned, the notion of knowledge is not under discussion here and is irrelevant to the current discussion. The idea of a contingent connection is supposed to explain how a description of the past is possible, given that nature is described from a point of view in the present.

It may be expected that, due to the fact that ascription of different truth-values to the same proposition was found to rely on a distinction between a proposition and what it describes, that is, between '*p*' and *p*, it is necessary to explain the ability to describe what exists. That is, as two independently existing entities in nature, it may be expected that a contingent connection between them is needed to explain the correlation between them. However, these expectations are based on our actual complex conceptual scheme, and it is not clear that they are relevant to the conceptual scheme I am developing here. I shall explore this suggestion when it arises naturally from considerations that relate to the conceptual scheme I am developing here.

Furthermore, I should emphasize the difference between the past and the future in this context. For I have shown that the distinction between present and past are necessary and fundamental distinctions in language, and hence must be used in non-inferential descriptions of nature. However, no such considerations apply to the notion of the future, which was shown to be definable based on the distinction between the past and the present. Hence, there is no need to assume that non-inferential descriptions of the future must be available, and no need to assume at this stage of the inquiry that there are contingent connections between future and present occurrences, analogous to those that were found necessary to explain the possibility of non-inferential descriptions of the past. That is, there is no need to assume that there are contingent connections between occurrences at different times, such that later occurrences are necessary conditions for earlier occurrences.

Moreover, I should emphasize that the explanation for the possibility of a non-inferential description of the past, that is, the existence of contingent connections between occurrences at different times, such that earlier occurrences are necessary conditions for the later occurrences, must be recognized in the any conceptual scheme. This is because this connection is required to explain and justify the conceptual structure that is included in this conceptual scheme. The coherence of this structure, therefore, requires that there is an available explanation for the possibility of non-inferential descriptions of the past.

Finally, I should stress the significance of this conclusion. For it may seem that it is trivial, and merely repeats what has already been known for generations, that is, that we rely on memory to maintain descriptions of the past (Augustine, 2019, p. 213). What is important in this conclusion is the necessary aspect of its conclusion, as is implied by the conceptual structure that is developed in Chap. 4.

5.6 Inferential Description of the Past

So far, I have argued that at least some of the basic propositions in language, which include the temporal indexical expressions "past" and "present," must be used non-inferentially to describe nature. In our rich conceptual scheme, however, some propositions about the past and the present are conclusions of inferences. The question is, is this possibility a necessary condition for the ability to formulate contingent propositions?

I have argued that to identify propositions as such, it is necessary to make a distinction between a proposition and what it describes. I further argued that this requires a notion of objectivity. This is reflected, for example, in the possibility of mistake. That is, every possible conceptual scheme that allows formulating contingent propositions must acknowledge the possibility that it ascribed wrong truth-values to its propositions. This recognition, as I have shown, applies to propositions that were once accepted as true but were in fact false, and vice versa.

Acknowledging the possibility of mistake raises a difficulty. It implies that at a certain time a proposition was ascribed a certain truth-value, while at a later time it is ascribed a different truth-value. Thus, for example, suppose that in the past in a conceptual scheme the proposition $P\varphi$ was considered true, and now it is regarded as false. If this is recorded in a straightforward manner in this conceptual scheme, this results in a contradiction. Furthermore, it is not clear which truth-value ascription is the correct one. How can this contradiction be avoided, and how can such conflicts be resolved?

The first problem, that of avoiding a contradiction, is solved by the ability to speak of propositions and their ascribed value. That is, while the past acceptance of the proposition Pφ as true is evaluated, this fact is not referred to as P(Pφ), or as P(T'Pφ'), but rather as P('T''Pφ'), that is, in the past the proposition Pφ was ascribed with the truth-value "true." Thus, it is possible to describe the conflict in the ascribed truth-values as N('(¬T)''Pφ') ∧ P('T''Pφ'). This possibility is important because it shows that the conceptual scheme which is developed here has the conceptual resources that are needed to undertake this inquiry. That is, it enables us to resolve such conflicts within our conceptual scheme without contradiction.

Interestingly, the same ability to resort to meta-linguistic terminology, which allows us to resolve this difficulty, is the one that gave rise to this difficulty. That is, the ability to distinguish between propositions and what they describe required a notion of objectivity, which gave rise to the possibility of mistake. The same distinction allows us to acknowledge and investigate this possibility consistently.

To resolve this conflict, it is important, firstly, to notice that such conflicts can only arise if they refer to at least one past ascription of truth-value to a proposition. That is, both ascriptions cannot be in the present, for this is a straightforward contradiction. Such conflict can only arise regarding a proposition that was once considered true, and at a later time, perhaps now, is considered to be false, or vice versa.

Next, I wish to distinguish between two types of possible mistakes. Notice that both types of fundamental propositions, propositions about the past, that is, Pφ, and about the present, that is, Nφ, imply the distinction between the proposition and what it describes, and hence are susceptible to the possibility of mistake. It might be thought that it is sufficient to establish the possibility of mistake regarding propositions about the present, that is, Nφ. This is because it can allegedly also explain the possibility of mistake regarding propositions about the past, that is, Pφ. After all, propositions about the past can be based, as I have shown in Sect. 5.5, on propositions about the present, that is, P(T'Nφ'). However, the possibilities of mistakes in both propositions are different. It can be shown that the possibility of mistakes regarding propositions of the type Nφ is not sufficient to explain the possibility of mistakes in propositions of the type Pφ.

To begin with, notice that a proposition of the type P(T'Nφ') makes a complex claim. It claims something about the conceptual scheme in the past, and about its relation to nature. Hence, two different types of mistake are possible, which should be carefully distinguished. The first is the possibility that the proposition Nφ was false *in relation* to the past, that is, the property φ was never realized in the past. The second is that, although Nφ was true *in relation* to the past, it was not *actually true* in this conceptual scheme in the past, because this conceptual scheme did not exist at that time—recall that this possibility was found necessary to explain the need for contingent propositions for describing nature. In other words, the first possible mistake arises due to the distinction between the conceptual scheme in the past and nature, while the second possible mistake arises due to the distinction between the conceptual scheme now and its state in the past. These are two gaps that give rise to two different possible mistakes—one is between the conceptual scheme and the world while the second is between the present and the past.

While proposition P(T'Nφ') is susceptible to two types of possible mistake, it is also possible to formulate propositions about the past that are susceptible only to the second type of mistake. Recall that it was shown in the previous chapter that a conceptual scheme must be able to describe its past ascriptions of truth-values, as demonstrated in the proposition P('T' 'Nφ'). Notice that in this case, the possibility of mistake relates only to the gap between the present and the past, and not to the gap between the conceptual scheme and the world.

I begin by exploring how the possibility of mistakes about the past can be identified and resolved. Recall in this context that the possibility of mistake can only be identified regarding propositions in the past (as explained in Sect. 5.2)—that is, it is only possible to determine that a proposition, whether about the future or the past, *was ascribed* a false truth-value. This implies that the gap between the present and the past is relevant for the identification of both types of mistake, that is, about the world and the past. Hence, it is the more fundamental type of possible mistake.

To begin with, the explanation, that was developed in Sect. 5.4, regarding the possibility of a non-inferential description of the past, cannot explain how it is possible to recognize the possibility of mistakes about the past. This can only explain how descriptions of the past can be available in the present. Such a mechanism is supposed to transform propositions that are held as true at one time to later times. Hence, it is difficult to see how it can give rise to conflicts between propositions. Furthermore, even if conflicting descriptions of the past are available, a criterion is needed to determine which description is right and which is wrong.

The idea of mistake hence requires another source of descriptions of the past, that is, it requires the possibility of non-inferential descriptions of the past. We are thus left with two questions. First, how can an alternative description of the past be acquired? And, assuming that there are conflicting descriptions of the past, how can it be decided which of these descriptions is correct? As shall soon become evident, the answer to both questions is the same.

I shall begin with the second question. Suppose that in a conceptual scheme a proposition of the type Pφ is found to be ascribed with conflicting truth-values. Only one of these truth-values is the correct one, and the difficulty is to determine which one. How can this be resolved? Obviously, a mere stipulation, for example, that the last ascription of truth-value ought to be accepted as correct is unjustified. The idea of mistake is inherently connected, as was shown, to the idea of objective truth. Hence, the question is, how can it be determined what happened in the past?

Although I do not offer a fully-fledged answer to this question in this section, it seems that the conclusion of the inquiry into the possibility of non-inferential descriptions of the past has paved the way for formulating an explanation. Recall that descriptions of the past are made possible by the existence of a contingent connection between past and present ascriptions of truth-values. Hence, the idea of the past is necessarily connected to the idea of contingent connections between occurrences at different times.

Hence, it seems that determining the truth-values of propositions about the past can rely on the existence of contingent connections between present and past occurrences. This allows formulating now inferential descriptions of the past, based on these contingent connections. Finally, this method of acquiring inferential descriptions of the past can also explain how conflicting descriptions of the past arise, and how they can be resolved.

However, the connections that were uncovered so far, between individual ascriptions of truth-values in a conceptual scheme, are unable by themselves to explain the possibility of inferential descriptions of the past. First and foremost, it requires expanding the idea of a contingent connection, which was so far identified with ascriptions of truth-values in the conceptual scheme, to nature in general. Furthermore, as will become apparent in what follows, it requires introducing the idea of regularity. This task will be undertaken in Sect. 5.8.

As a final note, I shall remark that inferential descriptions, once formulated, must be conserved non-inferentially over time. For, as I have argued, our conceptual scheme must be able to transfer descriptions from the past to the future, and the relation between these descriptions cannot be inferential. Nevertheless, I refer to any description that is preserved in a conceptual scheme as inferential if it was first accepted as true based on an inference.

5.7 The Asymmetry Between Past and Future

In the previous section, I argued that both non-inferential and inferential descriptions of the past must be possible. We may wonder, however, whether a symmetry between the past and the future is to be expected here.

Let me begin with the possibility of mistake. Recall that the possibility of mistake is important for recognizing a proposition as such, that is, as entities that are either true or false. Hence, is it also a necessary condition for propositions about the future. Can the conceptual resources that were uncovered so far explain how the possibility of mistake can be applied to propositions about the future?

One line of inquiry would therefore be to examine the possibility that a proposition about the future, that is $F\varphi$, which is accepted at one time as true, is recognized at a later time as false. This indeed seems possible, for I have argued that the notion of the future can be defined based on the distinction between past and present (as explained in Sect. 4.13). Hence, it seems that the objectivity of the future can be satisfied along the line that enabled introducing the notion of the future. Recall that the idea was to introduce the temporal indexical expression "future" as a relation between the past and the present. Let us suppose the property φ is realized now but was not realized in the past. In this case, it is conceivable that in the past the proposition $F\varphi$ was thought to be false, which is now recognized as a mistake. This seems sufficient for supporting the distinction between propositions about the future and their truth-value, and what they describe.

However, notice this scenario is not analogical to acknowledging a mistake regarding a proposition about the past. An analogical example would be a case in which it is now recognized that a proposition would be ascribed a wrong truth-value in the future in this conceptual scheme. In our conceptual scheme, this can be interpreted as a case in which I know now that I will have a false belief in the future. Although this may happen, this rarely happens in our conceptual scheme, and this fact highlights a fundamental difference in our conceptual scheme between the past and the future.

To better understand this asymmetry, recall that the distinctions between the past and the present were found to be the basic distinction, upon which the identification of the future relies. This implies that priority must be given both to the ability to formulate descriptions of the past and the present and to acknowledge the possibility of mistake regarding the past and the present, over making descriptions and acknowledging the possibility of mistake regarding the future.

Because the temporal basic distinctions are "past" and "present," and it is only possible to recognize a mistake about propositions in the past but not about the present (as explained in Sect. 5.2), the identification of mistake can only be made *after* the wrong ascription of truth-value to a proposition. This identification relies on an inferential description of the past, which itself relies on non-inferential descriptions as premises, and these inferential descriptions are preserved, thus expanding the description of nature in time.

The necessary conditions for giving meaning to the description of the past, and hence also for the possibility of mistake regarding the past, were thus shown to imply that our descriptions of nature are preserved and hence accumulating with the change from past to the present, or more generally, from earlier times to later times. Obviously, it is possible that some of the descriptions of a conceptual scheme would be lost—the connection that is responsible for connecting descriptions at different times is after all contingent. It is necessary, however, that some of the past descriptions are saved, and that new inferential descriptions are acquired and maintained. This process is hence inherently asymmetric between past and future.

Several questions arise at this stage regarding this asymmetry. For example, in our conceptual scheme, it is possible to formulate inferential descriptions of the future by relying on contingent regularities. In this respect, there seems that there a symmetry between the past and the future. However, while I have argued that the possibility of formulating inferential descriptions of the past is necessary to give meaning to the notion of the past, it is not clear that the same can be said for the possibility of formulating inferential descriptions of the future. As I have argued at the beginning of this section, the notion of the future is not basic, and relies on the notions of the past and the present. Hence, it may be that it is simply a matter of coincidence that our conceptual scheme allows for inferential descriptions of the future. In other words, it may be that it is a contingent property of nature, as we think about it, that it is characterized by contingent regularity that enables predicting the future.

On the other hand, what is missing in our conceptual scheme is the possibility of formulating non-inferential descriptions of the future. While I have argued that there must be a mechanism for preserving past descriptions, there is no such mechanism that relates to future descriptions. Again, it is not clear whether this is merely coincidental, or an indication of a conceptual impossibility.

I begin this analysis with the question of the inferential description of the future. This question is complicated and will be discussed in detail later. Currently, I would like to point out that, although in our conceptual scheme an inferential description of the future is possible, or seemingly possible, this possibility is not symmetrical in all aspects to the possibility of the inferential description of the past.

An inferential description of the future that is perfectly symmetric to the inferential description of the past must be of a reverse structure. An inferential description of the past is demonstrated in cases in which something existed in the past, without being described by the conceptual scheme at the time of its existence. Its description is acquired at a later time by inference and is preserved from that time on. An inferential description of the future that is symmetrical to an inferential description of the past would hence relate to something that exists in the future but is not acknowledged by the conceptual scheme at the time of its existence. It is learned about by inference at an earlier time, and is acknowledged before this inference, but not after it.

The latter description raises quite a few difficulties. For example, how can such a reverse inference be understood? Does it simply mean that the conclusion is accepted before the presentation of the premises, but not after it? Could it be described as an inferential description if it is accepted beforehand? Given our established ability to conserve truth-values of propositions from the past to the future, coupled with the fact that this description is accepted before the inference, it seems that describing this as an inferential description is unjustified.

The only way to accept this as an inferential description is if it is postulated that, similar to how descriptions of the past are preserved in time, there is a way of transferring descriptions from the future to the past. This ability, which is the reverse of memory in our conceptual scheme, however, seems to have far-reaching implications. This memory of the future can rely on contingent connections between the future and the past, such that the acceptance of a description in the future is a necessary condition for its acceptance in the past. Similarly, this mechanism can allow for the "accumulation" of more and more descriptions of future occurrences further in the past.

However, the most significant implication of this suggestion is that the symmetrical notion of an inferential description of the future depends on the possibility of a non-inferential description of the future. Recall that by a non-inferential description I do not mean an arbitrary ascription of truth-values to propositions. The possibility of non-inferential descriptions of the future depends on the possibility of a connection that allows transferring descriptions from the future to the past. In this context, recall that the problem of non-inferential descriptions of the past was solved by the ability to remember past descriptions of the (then) present, that is, $P(N\varphi)$. Similarly, once the idea of memory of the future is accepted, it follows that it is possible to remember future descriptions of (then) present, that is, $F(N\varphi)$. This implies non-inferential descriptions of the future.

The question is hence whether a non-inferential description of the future is indeed possible, for it is certainly not necessary. This is because, contrary to the notion of past and present, which are basic and hence rely on the possibility of non-inferential descriptions, the notion of future is supported by the notions of past and present, and hence does not require the possibility of non-inferential descriptions. Indeed, in our conceptual scheme, the possibility of a non-inferential description of the future does not seem to be available. The question that remains to be answered is, therefore: Is a non-inferential description of the future conceptually even possible?

Suppose that in a conceptual scheme, at some time a proposition Nφ is accepted as true, and this ascription of truth-value explains, via a contingent connection, the acceptance of the proposition (F(Nφ)) at earlier times, which hence counts as a non-inferential description of the future. While considering the possibility of such a suggestion, it should be noticed that if such a contingent connection does exist, according to which ascriptions of truth-values at later times explain the ascription of truth-values at earlier times, it could not be limited to propositions in the future which describe the (future) present. If such a connection exists, it should apply to any ascription of truth-value to any proposition in the future, because the content of the propositions and their origin are immaterial to the working of this mechanism. It should therefore apply to propositions about the (future) present, (future) future, and (future) past.

The implications of the possibility of a non-inferential description of the future are however far-reaching. I have argued that a necessary condition for the meaningfulness of propositions about the present and the past, which are the basic proposition in language, is the acknowledgment of their objectivity status, that is, acknowledging possibilities of ignorance and mistake, that rely on the possibility of learning new things about the past. This, in turn, was found to be based on an inferential description of the past. The problem is that the suggestion that we can learn new things about the past, via inferential descriptions, does not cohere with the possibility of a non-inferential description of the future. If a non-inferential description of the future is possible, there is no reason why it should not be available before it is "inferred" and hence nothing new is learned by this inference. Thus, the idea of learning new things about the past is undermined by the possibility of non-inferential descriptions of the future.

Recall that there is an asymmetry between the past and the future. A non-inferential description of the past, that is, a form of memory, is necessary for the coherence of a conceptual scheme that includes the distinction between past and present. A non-inferential description of the future, on the other hand, not only is not necessary, but also undermines one of the necessary conditions for the idea of objectivity of the past, and hence for the meaningfulness of the temporal indexical expression as a whole. Hence, if these ideas do not cohere, the idea of a non-inferential description of the past must be rejected.

It might be thought that the two ideas can coexist, under the assumption that both memory of the past and "memory of the future" are not perfect, and hence some of the non-inferential descriptions are forgotten. To begin with the idea of the memory of the past, admittedly, even though it is a necessary condition for the meaningfulness of the temporal indexical expressions, not everything is remembered. This imperfection, however, has nothing to do with the fact that memory of the future undermines the objectivity of the past. Another important point that should be noted in this context, is that even if the memory of the past would have been perfect, it would nevertheless leave open the possibility of learning new things about the past. This possibility, which is essential for the objectivity of the past, is undermined by the memory of the future.

Turning the attention to the suggestion that a memory of the future should likewise not be perfect, it might indeed seem that this possibility of ignorance about future states of the conceptual scheme, or "forgetting the future," leaves open the possibility of learning new things about the past, thus allowing for the objectivity of the past. That is, it leaves open the possibility of a case in which an inferential description of the past is acquired without memory of the future to anticipate this inference. However, the problem of contradiction is not solved by pointing out that there are cases in which there is no contradiction. The claim that a non-inferential description of the future is *possible* can only mean one thing—that it does not lead to a contradiction. However, it does lead to a contradiction, because it implies that it might be possible that a conceptual scheme both adopts a non-inferential description of the past while also rejecting this description as false due to memory of the future.

To understand where the suggestion of a non-inferential description of the future went wrong, I suggest looking back at Table 4.9 (in Sect. 4.13), in which I demonstrated how the order of occurrences in time is constituted by the basic indexical expression of past and present. What should be clear from this table is that (subjective) time is constituted by the accumulation of descriptions in time, that is, from the past to the present. This is an asymmetric process, which the possibility of non-inferential descriptions of the future undermines. Hence, non-inferential descriptions of the future are impossible.

There must therefore be a fundamental difference between the past and the future, which is connected to the idea of the flow of time. I have identified this asymmetry with the change in the truth-value of propositions in time, as this change is recorded in a conceptual scheme. What my analysis shows is that there is a necessary asymmetry between the past and the future, which is constituted by the fact that the foundation of the description of nature is found in propositions of the type $N\varphi \wedge P((\neg T)`N\varphi')$, which document the change in truth-value of propositions in the conceptual scheme, and which rely on the memory of the past. This is the basis for the idea that time flows: the memory of the past that allows for the documentation of the actual change in the truth-value of propositions in a conceptual scheme. The accumulation of descriptions in relation to the past explains why a past occurrence seems to be receding more and more from the present.

However, if a memory of the future was possible, then at any point in time a conceptual scheme would include propositions of the type $F(N\varphi)$, and the earlier the time, the more descriptions of the future are to be expected. This undermines the "flow of time." Again, similarly to forgetfulness of the past, it is possible that in earlier times fewer descriptions of the far future would be available. However, at each particular point in time, it would seem that the later time is receding, as if the present is moving towards the past. This is contradictory to the way we perceive the "flow of time." Hence, a memory of the future and a memory of the past are incompatible. My previous analysis showed that a conceptual scheme requires a change in the truth-values of propositions, and a record of this change in time. Now it is found that it must have a direction, and it cannot coherently occur in both directions.

One may nevertheless wonder whether it is not possible to consistently describe a conceptual scheme that includes both a memory of the past and a memory of the future, by separating between those that relate to past ascriptions of truth-values and those that relate to future ascriptions of truth-values. Thus, the accumulation of memories could be distinguished in two different directions. However, this suggestion seriously undermines the unity of the conceptual scheme by separating its contents. It implies separating the conceptual scheme into two different conceptual schemes, which might have the same conceptual structure, but have different content. If it is indeed a unified conceptual scheme, each "memory stream" must include descriptions of the content of its counterpart.

Finally, I should note that the conclusions of this analysis do not explain the asymmetry in time, but merely point out that it is a necessary feature of a conceptual scheme. Indeed, I should remind you that the temporal series of occurrences that I have constructed in Table 4.9 is of subjective time. I have yet to give meaning to the idea of objective time (a task which is undertaken in Sect. 6.3). Furthermore, the distinctions between past, present, and future are subjective, in the sense that they reflect the point of view of a conceptual scheme from which nature is described. Hence, the question remains to be answered, what explains the (objective) asymmetry in time? The explanation for this asymmetry, which is termed "the arrow of time," will be given in Sect. 7.5 of this book.

5.8 Laws of Nature and the Past Non-existence of the Conceptual Scheme

I have argued (in Sect. 4.11) that the temporal indexical expressions imply the possibility that the conceptual scheme did not exist in the past, and that this possibility must be acknowledged by this conceptual scheme. That is, the conceptual scheme must include the conceptual possibility that is demonstrated in proposition (d3) (which is a simplified version of proposition (d1)).

(d3) $P\varphi \wedge \neg P(T`N\varphi')$

According to proposition (d3), although in the past φ was realized, the proposition $N\varphi$ was never true in this conceptual scheme. Recall that this proposition claims that the proposition $N\varphi$ was never true, rather than that it was never ascribed with the truth-value "true." This is only possible if the conceptual scheme in which this description is formulated did not exist at the time in the past in which φ was realized. Thus, for example, although there were dinosaurs in the past, the proposition "now dinosaurs exist" was never true in my conceptual scheme.

As already mentioned, this conceptual possibility must be acknowledged as a conceptual possibility in this conceptual scheme, that is, it must be formulated and possibly accepted as true, to support the meaningfulness of the temporal indexical expressions. Hence, the conceptual scheme must have the conceptual resources that are necessary to identify the realization of this possibility. Otherwise, it would be useless for describing nature.

Accepting the conceptual possibility that is demonstrated in proposition (d3) requires the first conjunct in this proposition, that is, Pφ, to be an inferential description of the past. For if it was a non-inferential description, then the proposition Nφ was true in the past in this conceptual scheme, that is, P(T'Nφ'). However, this possibility is denied by the second conjunct in proposition (d3). Acknowledging this possibility hence requires the ability to infer the truth of a proposition about the past, that is, Pφ.

Answering this question is challenging, and in the present context, I am unable to solve all the difficulties that arise by trying to formulate a fully-fledged answer to this question. For example, notice that the proposition Pφ is general, as I explained in Sect. 4.10. That is, it claims that sometime in the past φ was realized. This raises the problem of how to settle this claim with the records of the past. Recall that the structure of the conceptual temporal dimension that was uncovered so far is demonstrated in Table 4.9. This structure allows a conceptual scheme to record a series of occurrences in time. Hence, to accept Pφ as true, it is necessary to determine when exactly φ was realized, that is, its location on the temporal series of occurrences. This in turn requires determining positions in time, and hence requires constituting time as an objective series of occurrences—a task that will be undertaken in Chap. 6.

Hence, I begin in this section by ignoring these complex issues, and limit the discussion to a simplified question, of how it is possible to infer that Pφ. I shall hence have to assume, as a working hypothesis, that this φ was never recorded as realized in this conceptual scheme. This is however only because at this stage of the inquiry, I have not introduced the conceptual means to determine an exact position in time. As shall become evident in what follows, I shall have to reject this working hypothesis. This, however, does not undermine my analysis, but merely exposes what is involved in making inferential descriptions of the past.

Since the proposition Pφ must be an inferential description of the past, the question that arises is: Which premises are required to justify this conclusion? To begin with, proposition Pφ is contingent, and hence can only be inferred from contingent propositions. Furthermore, I have already argued in Sect. 5.6 that inferential descriptions must ultimately rely on non-inferential descriptions. Finally, I have also suggested that the explanation for the possibility of non-inferential descriptions of the past suggests a way of explaining the possibility of inferential descriptions of the past. Recall that my analysis concluded that propositions (in the present) that describe the past non-inferentially must result from a contingent connection with past non-inferential descriptions of the (then) present. That is, there must be a contingent connection between past and present occurrences, such that would enable justifying the acceptance of inferential descriptions of the past.

However, the current challenge differs fundamentally from the previous discussion. It was previously established that there must be some connection between earlier and later occurrences, such that previous ascriptions of truth-value in a conceptual scheme are necessary conditions for later ascriptions of truth-value in a conceptual scheme. In the current context, however, the occurrences in question cannot be affirmations of propositions in a conceptual scheme, because no proposition which describes the realization of φ is presumed to be affirmed in proposition (d3). Furthermore, the connection between past and present ascription of truth-value is not rigorous. Not only does this connection permit exceptions, as is clear from our conceptual scheme, in which memory is fallible, but it must be so. This is because of the idea of objectivity, which requires descriptions of the past to be susceptible to the possibility of mistake. Hence, the conclusion of Sect. 5.5 is unable by itself to solve the problem I am facing now.

Hence the question is:, What are the necessary conditions for inferring that Pφ, under the assumption that the realization of φ was never recorded in this conceptual scheme? To begin with, to explain the ability to infer that Pφ, it is always possible to beg the question and simply rely on the premise that Pφ. This is, however, inappropriate in this context, in which the truth of the conclusion is important, rather than just the validity of the argument. What is required here is a premise, or a set of premises, *x,* which is connected by its truth-conditions to Pφ, such that, if *x* then Pφ. Furthermore, since this argument is formulated within the confines of a conceptual scheme that describes nature from a point of view that is located in the present, what is directly available to be used as a premise is located in the present, and hence is described by a proposition of the structure Nψ. Notice that this is a conceptual claim, which is based on the conclusions of Chap. 4, that applies to any type of proposition in language. That is, any proposition in language *y* is logically equivalent to N(T'*y*').

Hence, the argument which allows inferring the conclusion Pφ must be of the general structure (e):

(e1) Nψ
(e2) If Nψ then Pφ

Hence,

(e3) Pφ

Both premises of this argument must be contingent propositions. If Nψ is a necessary proposition, the second premise (e2) must also be a necessary proposition if it is true, and hence also the conclusion (e3). If the second premise (e2) is a necessary proposition while the first premise (e1) is a contingent proposition, the conclusion adds nothing to the meaning of the premise (e1), that is, Nψ, and hence this inference fails to extend the conceptual scheme's description of nature. It would be a description of the same fact in the present that is described by the first premise (e1). Hence, both premises of this argument must be contingent propositions.

It can therefore be concluded that the second premise of this argument, that is, "if $N\psi$ then $P\varphi$," must describe the existence of a contingent connection between two logically independent occurrences at different times, such that φ occurs before, and is a necessary condition for the occurrence of, ψ.

It is important to recognize the ambiguity in the proposition (e2), which can be interpreted in two ways. According to the first interpretation, proposition (e2) is a singular proposition, which asserts the existence of a contingent connection between two particular occurrences in time. According to the second interpretation, proposition (e2) is a general proposition, which describes the connection between types of occurrences, such that whenever an occurrence of a certain general type ψ occurs at a certain time, it is preceded by an occurrence of another general type φ.

It might be objected that the ambiguity I am identifying here is not justified by the conceptual apparatus that has been developed so far, but is rather imported from our complex conceptual scheme, which supports the distinction between token and types. In response, I should point out that this ambiguity is implied by the indexical expressions "past" and "present," as they were introduced into the conceptual scheme I am developing here. This is because these distinctions not only allow us to construct a series of occurrences in time, as I have previously shown. They also do not point to specific occurrences on this series of occurrences, and hence can apply, in different contexts, to multiple occurrences. For example, by using the already existing conceptual component of proposition (e2), it is possible to formulate a proposition in which two different sequences of occurrences of this type occur at different times, as demonstrated in proposition (33).

(33) $N(N\psi \wedge P\varphi) \wedge P(N\psi \wedge P\varphi)$

Hence, the ambiguity in proposition (e2), which is based on the distinction between a singular and a general proposition, is found in the confines of the conceptual scheme that was developed in the previous chapter, and is implied by the conceptual structure that was found necessary for the use of contingent propositions for describing nature.

Once the distinction between the two interpretations is brought to light, it is obvious that although both interpretations allow deducing the conclusion $P\varphi$, only the interpretation of proposition (e2) as a general proposition can explain the possibility of acquiring new inferential descriptions of the past. For to accept the conclusion as true, the argument should not only be valid, but also sound, and should not beg the question. To rely on this argument, it is hence necessary to accept the premises as true. However, as a singular conditional proposition, and given the truth of the antecedent, proposition (e2) can only be accepted as true if the consequent is also accepted as true. However, the consequent of proposition (e2) is the conclusion of the arguments, and hence such interpretation of (e2) begs the question.

Proposition (e2) must be therefore interpreted as a general proposition, which asserts the existence of a contingent connection not merely between individual occurrences, but between types of occurrences. That is, to explain the possibility of acknowledging the non-existence of the conceptual scheme in the past, as demonstrated in proposition (d3), proposition (e2) must be interpreted as describing a contingent general connection, which ties together types of occurrences at different times, such that the later occurrences are sufficient condition for the occurrence of the earlier occurrences.

As a general contingent proposition, proposition (e2) must be established empirically. This is possible if instances of this general proposition are identified by this conceptual scheme in the past, that is, if it was recognized before that φ is followed by ψ. This, under the assumption that nature is necessarily subject to contingent regularity, justifies adopting premise (e2) based on descriptions of its past instances. Once affirmed, it is possible to rely on proposition (e2) as a premise to infer, in this particular case, the truth of the conclusion (e3).

The conclusion of this analysis is hence that only the existence of contingent regular connections between types of occurrences can explain the possibility of acknowledging the non-existence of the conceptual scheme in the past. This implies that the existence of such contingent regular connections is a necessary feature of nature, and must be acknowledged by any conceptual scheme that has the conceptual resources for formulating contingent propositions.

It is important to recognize the difference between this conclusion and the conclusion of Sect. 5.5, regarding the necessary conditions for formulating non-inferential descriptions of the past. The conclusion of Sect. 5.5 refers to a contingent connection that obtains between individual occurrences. Although it might be argued that this conclusion implies a multitude of such connections, they always seem to be singular connections. Here, on the other hand, the conclusion is the existence of a contingent regular connection that holds between types of occurrences.

Furthermore, the conclusion of Sect. 5.5 fell far short of establishing the existence of a lawful connection between occurrences in time. This is because all I have argued for is that there must be a contingent connection between *some* occurrences at different times. Moreover, and as mentioned earlier, even if the conclusion of Sect. 5.5 could be interpreted as supporting a contingent regularity, it cannot be interpreted as supporting a strict law. Regularities that tie together ascriptions of truth-values in the present and past must permit exceptions, because they must allow for the possibility of mistake. These regularities, which obtain between ascriptions of truth-values, can be explained by being supported by true laws of nature, which allow for no exceptions, but do not constitute laws of nature themselves.

Turning to our conceptual scheme, it is easy to find examples of contingent regularities that possibly can be, and sometimes actually are, violated. Examples of such regularities are memory, the fact that the earth supports us, that the bread we eat is nourishing, that the sun rises and sets regularly, and so on. These regularities are explained by their reliance on true laws of nature, that is, rigorous general rules that allow no exceptions.

I should note in this context, that some philosophers have doubted whether there are exceptionless regularities in nature (Cartwright, 1980; Giere, 1999; Lange, 1993). However, in the context of the present analysis, this claim is unacceptable. This is because the question of the existence of the laws of nature applies to the objective realm, which is necessarily constituted by the laws of nature (see also (Yehezkel, 2014, pp. 121–135)).

Proposition (e2) represents a true law of nature, such that it allows for no exceptions. This is because, as I have argued, it is only possible to describe the past from a point of view that is located in the present. Coupled with the possibilities of mistake and ignorance, this implies that an objective description of the past is based on a general regularity that connects present occurrences and past occurrences, such that past occurrences are necessary conditions for present occurrences. This regularity must be rigorous and allow for no exceptions because the notions of mistake and ignorance require a criterion for truth. Inferential descriptions rely on this criterion. Since there is no way to describe nature other than from the present, and given this is only possible by relying on a regularity of the type that is demonstrated in proposition (e2), only this regularity can determine what happened in the past. Hence, it can never be found violated. It is itself the ultimate criterion for truth.

I now reiterate and summarize the analysis of this section. The idea of the past requires the ability to determine, from the present, that φ was realized in the past. This is only possible by relying on a contingent premise of the type "if Nψ then Pφ." The premise "if Nψ then Pφ" can only be a general proposition, in order not to beg the question. As a general proposition, not only is it possible to endorse it without examining every singular case that falls under it, but it is also necessary to do so, because of the necessarily limited point of view of nature, which is implied by the notion of objectivity.

Notice that this conclusion does not raise the problem of induction, and in fact offers a partial answer to this problem. It is now clear that Hume was mistaken in his claim that the "Uniformity of Nature," according to which "the course of nature continues always uniformly the same," is not a conceptual truth, or "relation of ideas" (Hume, 1978, p. 89). Hume argued that it is possible to "conceive a change in the course of nature." However, it is now clear that Kant was right, and the very notion of objectivity, that allows general propositions to extend beyond our limited experience of nature, is constituted by laws of nature (Kant, 1929, pp. B218–B265). However, I should emphasize that it is only a partial answer to Hume's challenge, and I shall return to this problem in what follows (in Sect. 5.10).

Although this regularity cannot obtain, as I have explained, between ascriptions of truth-values in a conceptual scheme, there must be occurrences in nature that do fulfill this requirement. Furthermore, since the indexical expressions of past and future do not refer to specific times, this conclusion is general, and connects occurrences in all times. That is, in each time there must be occurrences whose existence is sufficient for determining past occurrences.

The notion of time hence relies on the idea of laws of nature, which tie together occurrences at different times, such that later occurrences are sufficient conditions for the existence of previous occurrences. While it is a necessary feature of nature that it is susceptible to such laws, the specific laws themselves are contingent. This seemingly paradoxical conclusion will be further analyzed and explained in what follows.

5.9 Laws of Nature and the Symmetry Between the Past and the Future

I have argued that the concept of the past requires objectivity, that is, a distinction between what is described by the conceptual scheme and what is actually the case. To apply this distinction, it was found necessary to infer how things were in the past based on the way things are in the present. For this purpose, it suffices that the occurrences in the present that are available as a basis for this inference are a sufficient condition for the past occurrences. That is, this inference relies on laws of nature, according to which later occurrences of a certain type are a sufficient condition for earlier occurrences of a certain type.

The question I wish to address now is as follows. Is a symmetry required between the past and the future, such that earlier occurrences are also a sufficient condition for later occurrences?

A first suggestion for supporting a positive answer to this question relies on the fact that the future must also be objective, and hence independent of its description from the conceptual scheme in the present. Thus, propositions about the future imply the possibilities of mistake and ignorance just as do propositions about the past. Hence, it seems that anything that applies to the past, in terms of the necessary conditions for securing its objectivity, must also apply to the future.

This suggestion, however, raises several difficulties. To begin with, to explain the possibilities of mistake and ignorance regarding the future, there seems to be no need for inferential descriptions. For example, while I do not know now what will happen in the future, if I wait long enough, I can find this out through non-inferential descriptions. That is, I can find the answer by witnessing and remembering what will happen. Further, notice that this is the same for fulfilling the necessary conditions for the objectivity of the present.

Furthermore, as I have argued, the concept of the future is not as fundamental as the concepts of the past and present, and can be defined based on these concepts. Hence, it seems that the concept of the future does not imply any special conditions for its inclusion in a conceptual scheme, beyond those that are necessary for giving meaning to the concepts of the past and present.

It hence seems that inferential descriptions of the past, which rely on the premise that later occurrences are sufficient conditions for earlier occurrences, suffice for giving meaning to the distinctions between past, present, and future. However, although I have argued that the laws of nature are sufficient for giving meaning to inferential descriptions of the past, I have yet to explain how such laws are identified. More specifically, I have yet to explain how it is possible to determine that occurrences of some type are sufficient conditions for earlier occurrences of another type. Recall that this is a necessary condition for formulating inferential descriptions of the past.

Suppose that the occurrence of a past occurrence of type φ needs to be determined. Allegedly, all that is needed for making such objective determination possible is to stipulate that it is tied with at least one later occurrence of type ψ, such that ψ's occur only if φ's occur before them. Hence, if an occurrence of type ψ occurs, it is possible to infer the existence of a previous occurrence φ.

However, notice that this connection between later and earlier occurrences does not allow us to determine objectively that φ did not exist under any circumstances. If ψ occurs, it is possible to conclude that φ occurred in the past. However, if ψ does not occur, it is impossible to conclude that φ did not occur. Hence, it is impossible to determine the non-existence of φ. Furthermore, if ψ does not occur, the existence of φ remains undecided, as it is impossible to determine both that it existed and that it did not exist.

I should emphasize the meaning of this implication. It implies that there is never a reason for inferring that φ did not occur, and hence it could be always assumed that φ exists, not only in the past but at all times. Thus, the distinction between the existence and non-existence of φ is lost, and hence, according to the conception of language as a system of coordinates, the concept of existence loses its meaning, as it fails to make any contribution to descriptions of nature.

Further notice that the inability to determine non-existence also implies that it is impossible to explain the possibility of mistake, that is, the possibility that φ did not exist although a description of its existence was affirmed in this conceptual scheme. I have argued that the possibility of mistake is essential for the distinction between a proposition and what it describes, and hence for the idea of objectivity. The suggestion that past occurrences are only necessary conditions, but not sufficient, for future occurrences hence undermines the idea of objectivity, which is itself necessary for explaining the need for contingent propositions for describing nature.

Can the difficulty of determining non-existence be resolved without postulating that earlier occurrences are sufficient conditions for later occurrences? To answer this question, I shall try to identify the necessary conditions for determining φ's non-existence, without relying on the premise that φ sufficiently determines ψ.

To determine that φ did not exist from the present, it would be necessary to connect φ's non-existence with some later occurrence other than ψ, for example, μ, such that μ's existence implies that φ did not exist. From this connection, it follows that if φ existed, then μ did not exist. That is, φ's existence is a sufficient condition for μ's future non-existence. However, according to this stipulation φ's existence is not a sufficient condition for ψ's existence, which may not exist even if φ existed. Hence, if both ψ and μ do not exist, it is impossible to determine whether φ existed in the past or not.

I should emphasize the gravity of the impossibility of giving a definite answer to the question of whether something objectively existed in the past or not. Admittedly, it may be practically impossible to determine what happened in the past, as is many times the case in our conceptual scheme. Thus, for example, it is practically impossible to determine whether a humanoid sneezed on the very spot in which I am now exactly 10,000 years ago. However, a conceptual impossibility to accurately determine what happened in the past contradicts the very idea of objectivity that was found necessary to explain the flow of time, which itself is necessary for the contingency of propositions in language. That is, my analysis has shown that nature must be described from a point of view in the present, and that an objective view of the past is a necessary condition for existence. This objectivity is constituted by the laws of nature, and the laws of nature must allow determining, from the point of view that is located in the present, what existed in the past. The suggestion that it is impossible to objectively determine in principle whether something existed or not is a conceptual contradiction, which undermines the complex structure that I have constituted in the previous chapter.

Turning back to our previous example, this conclusion implies that the existence of ψ and μ cannot be independent of one another. That is, it must be a nomological impossibility for ψ and μ not to exist simultaneously, as is it impossible for them to exist simultaneously. Hence, φ existed in the past if and only if ψ exists now and μ does not exist; and φ did not exist in the past if and only if μ exists now and ψ does not exist. This is the only way to give meaning to the objective existence of φ in the past.

My analysis concludes that every occurrence in time must be a necessary and sufficient condition for later occurrences. These later occurrences must also be similarly connected to subsequent occurrences. Thus, the laws of nature create continuous chains of interconnected occurrences in time.

I should note, however, the limitations of this conclusion. To begin with, this conclusion does not preclude the possibility that these chains of occurrences will be completely independent of one another, such that no occurrence in one chain is connected with occurrences in other chains. Furthermore, although this conclusion precludes these chains of occurrences from ever coming to an end, it does not rule out the possibility of an occurrence suddenly appearing without being connected to any previous occurrence. This is because the past is determined from a later point of view, that is, the present. Chains of occurrences that end, however, cannot be identified from a later point of view. Prima facie, this does not rule out the possibility of a beginning of a chain of occurrences. This possibility will be discussed in Sect. 5.11.

5.10 Determinism of the Future and the Problem of Induction

I have argued in the previous section that past occurrences must be a sufficient condition for later occurrences. However, it might be argued that this does not imply determinism of the future, that is, it does not imply that a description of the past coupled with the laws of nature determine only one possible description of the future. It might be, it can be argued, that the past determines that one of several possibilities will occur in the future. That is, rather than the previous occurrence of φ determining that ψ will occur later, φ implies, for example, that either ψ or μ will occur. According to this suggestion, the laws of nature are hence not deterministic.

The question of whether the laws of nature are deterministic is not epistemological, but rather metaphysical. That is, I do not inquire whether it is possible to know the future in advance. I have discussed the asymmetry in the ability to describe the present and the future in Sect. 5.7, where I argued that there must be an asymmetry in this respect between the past and the future. This, however, does not imply that the future is not determined by previous occurrences, in symmetry to how the past is determined by later occurrences. An epistemological asymmetry does not imply a metaphysical asymmetry that is reflected in the structure of the laws of nature.

Nevertheless, the suggestion that the future is not determined can be associated with the said asymmetry between the past and the future, which was found to be a necessary feature of any conceptual scheme. According to this asymmetry, the past must be determined from a later point of view of nature, because any conceptual scheme must describe nature from the present. This distinction between the past and the present is the basis of the contingent description of nature, and allows giving meaning to the idea of the future. Hence, although it is necessary to rely on the premise that the past is determined by later occurrences, prima facie the future is not subject to this stipulation, and can allegedly be given meaning without relying on the premise that future occurrences are determined from a point of view in the present.

Furthermore, it might be argued that while the past must be determined objectively from the present based on laws of nature, the application of the idea of objectivity to the future does not seem to require an analogical conceptual apparatus. Applying the distinction between the ascription of truth-value to a proposition about the future and what will actually be the case in nature can rely on the accumulation of the non-inferential descriptions of the past. For example, while I thought when I was young that I would study physics, I was wrong and in fact studied metaphysics. To identify this mistake, it suffices to wait until the future occurrences become present, there is no need to rely on laws of nature to predict the future. Furthermore, with time, things that were in the future became past, and can be determined objectively by relying on the determination of the past by later occurrences. Again, it seems that there is no need for postulating laws of nature according to which the future is determined by past occurrences.

Notwithstanding all these considerations, it can be shown that the future must be determined by previous occurrences. To understand why, consider again the suggestion that the occurrence of φ is a sufficient condition for a disjunction of ψ and μ, but not for either of them separately. This allows inferring the existence of φ from the existence of each of them separately, as well as the non-existence of φ from the non-existence of both occurrences, ψ and μ. Thus, it is allegedly possible to give meaning to the objectivity of the past without relying on the premise that the future is determined by past occurrences.

The previous suggestion relies on the premise that it is possible to establish that every occurrence of φ is followed by either ψ or μ. However, how can it be determined that this is indeed the case? What would distinguish, for example, between ψ and μ which are connected to φ by the laws of nature, on the one hand, and another occurrence ξ that happens to occur after φ, but is not linked to the previous occurrence of φ by the laws of nature? That is, how is it possible to determine that only ψ and μ should be included in this disjunction, rather than any other occurrence, or indeed all occurrences, that takes place after each occurrence of φ? Furthermore, how can a claim for the existence of such a law of nature be disproved, if in each case in which φ occurs, but is not followed by ψ or μ, it is possible to arbitrarily identify another occurrence ξ, and add it to the disjunction of occurrences that are allegedly linked to the previous occurrence of φ? All these questions are unanswerable under the assumption that the laws of nature are indeterministic, and show that it is impossible to establish the existence of contingent laws of nature, and hence also to rely on such laws for determining objective existence. Indeed, they show that the idea of indeterministic laws of nature is vacuous, as such laws can be made to fit a completely chaotic series of occurrences.

Notice that the difficulties I am raising here do not apply to deterministic laws of nature, that is, laws that stipulate that the same occurrences will always be followed by the same occurrences. Although there may be difficulty in identifying the specific contingent connection that obtains between occurrences at different times, the fact that these laws stipulate that the same circumstances will always be followed by the same occurrences offers a way of testing any suggested law of nature.

I should emphasize, however, that the problem with the idea of indeterministic laws of nature is not merely epistemological, and my criticism does not amount to the claim that such laws would be impossible to verify or even support. Recall that the laws of nature were found to be necessary to determine what objectively exists, and their existence is essential for explaining the need for contingent propositions for describing nature. Hence, these laws are constitutive of the objective realm, that is, of nature. Thus, the problem of identifying the laws of nature is not raised on the backdrop of an independent objective conception of nature, according to which nature is independent of the conceptual scheme from which it is described. It is in fact the other way around. The objectivity of nature is constituted by these laws of nature, which must therefore be identifiable by any conceptual scheme from which it is described.

It is important to recognize the fact that the idea of indeterministic laws of nature undermines their ability not only to identify what exists, although this is the reason why the idea of laws of nature was introduced. It also undermines their ability to determine what exists, although they are supposed to constitute the objective realm.

To see why, recall that the general regularity that was found to be necessary for explaining objectivity is contingent, and hence must be based on non-inferential descriptions of nature. Admittedly the laws of nature cannot be verified, not only because of their general nature, but also because they were introduced exactly to allow for extending the description of nature beyond the non-inferential descriptions of nature. As general laws they should, however, be confirmable, that is, supported by their instances.

Suppose that an instance of φ followed by ψ is identified. Given the assumption that nature is subject to a general regularity (which I have established), it is justified to infer that, all things being equal, if another φ occurs, it will be followed by an instance of ψ. Admittedly, there is a possibility of mistake, because of the difficulty of establishing the relevant factors, and the possibility that the properties ψ and φ, even if they exhibit some kind of regularity, are not fundamental and the connection between them does not constitute a law of nature. These possibilities of mistake, however, relate to the identification of specific laws of nature, but not to the existence of these laws.

The justification for extending beyond the finite number of instances of which there are non-inferential descriptions, by generalizing for all that exists in nature, is the assumption that nature is indeed subject to regularity. This regularity implies that everything in nature is similar to what is available by non-inferential descriptions. This assumption justified extending non-inferential descriptions to the past, present, and future, with the help of inferential descriptions. Without this assumption, according to which in nature similar occurrences are followed by similar occurrences, it is not justified to rely on non-inferential descriptions to formulate inferential descriptions. In terms of our conceptual scheme, it is justified to draw conclusions from what is experienced to what is not experienced only under the assumption that under similar circumstances similar occurrences will follow. For example, if I know that the change in colors on a TV screen is governed by chance, is it unjustified to infer from one instance in which a blue screen was followed by a white screen that this will also be the case after the next appearance of a blue screen?

The general premise that underlies any generalization, and allows the identification of laws of nature, is hence that in similar circumstances the same occurrences will follow. However, now it is possible to see that the idea of indeterministic laws of nature is self-contradictory. For it states that even if all the relevant circumstances are reproduced, or indeed that everything that existed in nature is reproduced, the same occurrences need not follow. Hence, it undermines the very idea that justifies the generalization, that is, the idea of laws of nature. "Indeterministic laws of nature" are not regularities at all, and hence this term is self-contradictory.

I might insist that an indeterministic law of nature is nevertheless a regularity, simply not a deterministic regularity. Instead of claiming that φ is always followed by ψ, it is claimed that after φ either ψ or μ follows. Furthermore, this can be justified by observation, just as deterministic regularity is identified.

Notice, however, the fundamental difference between deterministic regularity and the alleged indeterministic regularity. Deterministic regularity requires, in principle, only one instance. Further instances are needed only because of the possibility of misidentifying the relevant factors, and to rule out the possibility that the properties that are identified are not the fundamental properties that are susceptible to the laws of nature. Hence, even without identifying the relevant factors and the fundamental properties of nature, it implies that if everything in nature were to be reproduced exactly as was observed, the same occurrences would follow.

In the case of "indeterministic regularity," on the other hand, it is impossible to formulate a law of nature based on one observed instance, in which either only ψ or only μ occur after φ. Furthermore, if one observed instance will not do, adding more observed instances would not do either. Experiences in which ψ or μ follow φ can teach nothing about what happens after the next instance of φ. This inference relies on the assumption that under the same conditions the same occurrences will follow. However, if this does not hold in the case of the rule "if φ then ψ," neither does it hold in the case of "if φ then ψ or μ." If it is impossible to infer, based on one experience of φ followed by ψ, what would happen the next time φ occurs, it is also impossible to infer based on several experiences what would happen the next time φ occurs. If the fact that ψ occurred once after φ does not imply that the same would happen the next time φ occurs, why would the fact that either ψ or μ occurred after φ's imply anything about what would happen after the next occurrence of φ? Once the principle, that under the same circumstances the same occurrences will follow, is undermined, so is any attempt to project from part of nature to all of nature, no matter how large or small this part is. This is Hume's point in his claim that any attempt to conclude from what is experienced to what is not experienced relies on the principle of the uniformity of nature (Hume, 1978, p. 89).

Rather than suggest a different type of laws of nature, the idea of indeterministic laws of nature amounts to a rejection of laws of nature. This does not imply the rejection of statistical laws. It does imply the rejection of their status as laws of nature. Statistical laws can only rely on deterministic laws, and only deterministic laws can be laws of nature.

It can therefore be concluded that the laws of nature that are needed to establish an objective conception of nature must be deterministic, and therefore not only must the past be determined by later occurrences, but the future must also be determined by earlier occurrences. Time is hence symmetrical in terms of the structure of the laws of nature.

Before concluding this inquiry, two explanatory remarks are in order. First, I wish to clarify the notion of necessity involved in this type of determinism. Since it is the outcome of a conceptual analysis, it is best described as a *conceptual necessity* rather than a *logical necessity* (which, for example, is associated with "logical determinism") or *metaphysical determinism* (which may be based on general metaphysical or ontological considerations). This type of determinism further implies *nomological determinism*, that is, determinism associated with causality or contingent laws of nature. This raises questions about the dual aspect of the laws of nature, which have both a necessary and a contingent dimension. This complex issue requires a separate discussion, which will be undertaken in Sect. 5.15.

The second remark concerns a possible conflict with certain interpretations of Quantum Mechanics, according to which nature is fundamentally random and cannot be fully described by deterministic laws. In response to this concern, I should first remind the reader that my aim in this book is to establish natural philosophy as an independent and a priori science. *Natural Philosophy*, in this sense, is a complementary research project to physics rather than a branch of the philosophy of science that follows in the footsteps of empirical natural science. To establish this discipline and secure the rigor of its conclusions, it is necessary to conduct this research on a priori grounds alone. Although the conclusions of both endeavors should ideally align, it is not yet clear that they have reached this point.

Therefore, I will be content here to offer a way in which the conclusion of my analysis might coherently coexist with quantum mechanics. The analysis of quantum mechanics exceeds the scope of this inquiry; however, I suggest that the apparent tension may lie in the possibility of *epistemological uncertainty*. In other words, it may be that the seeming indeterminism allegedly involved in quantum mechanics reflects an inherent inability to acquire complete information about distinct parts of nature. This suggestion is further supported by considerations on the meaning of the laws of nature, as discussed in Sect. 7.3. Thus, the conclusion of my analysis incorporates determinism with epistemological uncertainty.

5.11 The Unity of Nature

I argued that the laws of nature connect occurrences at different times. Because the point of view from which nature is described is the present, and given the requirement for an objective conception of the past, each occurrence must be connected to at least one later occurrence, which in turn needs to be connected to another occurrence, and so on. This creates a chain of occurrences that are connected by the laws of nature.

The fact that all occurrences must be part of such chains does not imply that all occurrences are connected. The conclusions of my analysis so far have left open the possibility that there are several chains of occurrences that do not connect. This is the possibility I wish to address now.

I have argued that objectivity, as demonstrated in the possibilities of mistake and ignorance regarding propositions about the present, is reflected in the distinction between the ascription of truth-value to propositions and what is actually the case. This distinction implies a coexistence of different entities in nature, that is, the proposition and what this proposition describes. Nature must hence allow for different entities to exist at the same time.

Given that two entities exist at the same time, according to the previous conclusion, each entity must be part of a chain of occurrences that are connected by the laws of nature. Each of these chains connects occurrences at different times. Hence, so far, the conclusions of my analyses failed to find and connect between two coexisting entities.

The question that arises is, can the different chains of occurrences, in which occurrences are connected by laws of nature, be completely independent of one another, that is, not tied at all to one another by the laws of nature? Furthermore, if this is indeed possible, can such chains have a beginning, that is, can some occurrences not be tied by the laws of nature to previous occurrences? Notice that these questions are connected because, as will be explained in what follows, a negative answer to the former question will impose a general regularity on nature, which would limit the possibility of occurrences whose existence is not tied to laws of nature.

To answer the former question first, recall that describing nature is undertaken from the point of view of a conceptual scheme. Thus, the identification of any occurrence, as well as the identification of the laws of nature that obtain between occurrences, must be explained by the way that the laws of nature connect the occurrences in nature to the conceptual scheme, from whose point of view these occurrences are described.

Admittedly, as explained before, the connections between occurrences and the ascription of truth-values to propositions cannot constitute laws of nature, although they do rely on the laws of nature. This is because the laws of nature allow for no exceptions, and hence cannot explain the possibilities of mistake and ignorance. These possibilities are explained by natural regularities which rely on the laws of nature, but are not themselves laws of nature. This only implies that the ascription of truth-values is not a fundamental feature of nature, and it does not undermine the conclusion that everything in nature is ultimately subject to the laws of nature.

To reiterate, existence in nature is determined from the point of view of a conceptual scheme. This existence is objectively determined by the laws of nature, which tie together occurrences at different times. Thus, the laws of nature enable the conceptual scheme to describe nature, whether by non-inferential or inferential descriptions of what exists. Although the connections between the descriptions and what they describe do not constitute laws of nature, they imply that all the occurrences that are described by the laws of nature are interconnected by the laws of nature at least at one point—the conceptual scheme from whose point of view nature is described. This conceptual scheme, however, cannot be the only place in which the chains that are created by the laws of nature intersect, because the conceptual scheme only supervenes on the laws of nature. As previously mentioned, the regularities that connect occurrences to the ascriptions of truth-values in the conceptual scheme do not constitute laws of nature, but do rely on the laws of nature. Hence, the conceptual scheme does not demonstrate any exceptional function of the laws of nature, nor does it occupy a distinctive position in the structure of the laws of nature. If different chains of occurrences in nature interconnect in the conceptual scheme, they must be able to interconnect as a general rule.

It can therefore be concluded that it is impossible for separate chains of occurrences, which are independent of one another, to be part of nature. This is because all chains of occurrences must all be described together by a single description if they are to be recognized as existing together. Independent chains of occurrences could not be recognized as part of nature, or even as existing independently. As Kant has argued in the Transcendental Deduction of the Categories (Kant, 1929, pp. B132–B136), the idea of multiple existing things requires their unity in a single description. Hence, the unity of nature requires the unity of the conceptual scheme, and both are made possible by the laws of nature, which tie together all occurrences in nature.

In light of this conclusion, it is clear that occurrences cannot begin to exist without any connection to previous occurrences. This is because, for all the chain of occurrences to connect, the laws of nature cannot be limited to the connection between two different and isolated occurrences at different times. On the contrary, they must connect different coexisting occurrences with other occurrences at different times. Suppose that an occurrence φ could begin to exist without being determined by any previous set of occurrences ψ. In this case, any later occurrence μ, which is determined by φ, cannot be determined by any of the previous set of occurrences ψ. This would imply a violation of the principle that earlier occurrences must be sufficient conditions for later occurrences, thus violating the laws of nature.

In conclusion, because nature is described from a point of view of a conceptual scheme, which is distinguished from what it describes, nature must be subject to laws. These laws must connect all the occurrences in nature, thus allowing formulation of an objective description of nature from a single point of view of nature, that is, the point of view that is constituted by the conceptual scheme. Again, I note that this does not represent a unique position of the conceptual scheme in nature. It simply reflects the fact that what does not connect to the rest of nature cannot be considered a part of it, and hence cannot be considered to exist at all. Existence, as I have previously explained, is relative.

5.12 The Impossibility of Simultaneous Laws of Nature

I have argued that nature must be subject to laws that connect occurrences at different times. These laws of nature are first and foremost supposed to explain the ability to formulate objective descriptions of the past from a point of view in the present, and hence it is necessary for the contingent connections they describe to obtain between occurrences at different times. However, the reader may wonder whether there is also room, or perhaps even a need, for simultaneous laws of nature, that is, laws of nature that connect simultaneous occurrences by necessary and sufficient conditions. Such proposed laws of nature determine that given an occurrence φ at a certain time, another occurrence ψ must exist simultaneously.

I wish to emphasize the difference between the laws of nature that I have introduced in the previous sections, which in the current discussion will be referred to as non-simultaneous laws of nature, and the simultaneous laws of nature whose possibility is discussed here. Non-simultaneous laws of nature are those that rely on time, and hence cannot be applied to an atemporal realm. The temporal aspect of these laws is reflected by the fact that the regularity that they describe holds between occurrences at different times. Simultaneous laws of nature, on the other hand, can apply to occurrence in an atemporal realm. For example, a law that describes the structure of a chessboard-like realm, according to which a white square is always next to a black square, in both row and column, and vice versa.

My main interest in this section is to explore the possibility of simultaneous laws of nature. However, it might be thought that such laws are not only possible, but may even be necessary. This is because objectivity is equally required for propositions that describe present occurrences; that is, there must be a distinction between the truth-value that is ascribed to propositions about the present and what is actually the case. This implies that there must be things that exist now but are not described by the conceptual scheme. Because the objectivity of the past was found to be necessarily explained by non-simultaneous laws of nature, should the objectivity of the present be explained with the help of simultaneous laws of nature?

Admittedly, the ability to apply to propositions about the present the distinction between what is described by the conceptual scheme and what is actually the case is fulfilled by the necessary conditions for the objectivity of the past. For what is now present will become past, and hence it is possible to determine now that in the past the conceptual scheme did not accurately describe what existed at that time. That is, it is possible to acknowledge the falsehood of propositions about the present by applying the distinction between the acknowledged truth-value and the actual truth-value to propositions of the type P(Nφ), as was explained in the previous sections. It hence seems that all the necessary conditions for giving meaning to the objectivity of the present are fulfilled by relying on laws of nature that connect occurrences at different times.

However, it might be argued that this fact, which shows that objectivity of the present can rely on the objectivity of the past, does not prove that this is necessarily the case. Perhaps, it might be argued, it is possible to give meaning to the objectivity of both the past and the present by relying on simultaneous laws of nature. Furthermore, it might be argued that even though simultaneous laws of nature are not necessary, it does not imply that such laws do not obtain as a matter of fact.

To begin with, I should emphasize that laws of nature that connect occurrences at different times were proved to be a necessary feature of nature, and hence fulfill a role that no simultaneous laws of nature can replace. This is because the recognition of the possible non-existence of the conceptual scheme, which was found necessary for explaining the need for contingent propositions for describing nature, requires acknowledging the existence of times in which the conceptual scheme did not exist. This is only possible by relying on laws of nature that connect occurrences at the time of the existence of the conceptual scheme with occurrences at the time in which the conceptual scheme did exist. Furthermore, both inferential and non-inferential descriptions of the past were shown to rely on laws of nature that connect occurrences at different times. Since the notion of the past is basic, and cannot be defined with the help of the notion of the present, the non-simultaneous laws of nature are irreplaceable.

Hence, in light of the conclusion that non-simultaneous laws of nature are a necessary feature of nature, the question of the possibility of simultaneous laws of nature is first and foremost determined by the question: Can simultaneous laws of nature coexist with non-simultaneous laws of nature?

To begin with, it should be pointed out that the idea of simultaneous laws of nature is problematic, due precisely to its claim to give meaning to the idea of the objectivity of the present. This is because simultaneous laws of nature allegedly allow inferring the existence of present occurrences. However, the idea of objectivity relies on the distinction between the truth-value that is ascribed to a proposition and its actual truth-value, and the laws of nature were introduced precisely for giving meaning to this distinction. However, this use of the (non-simultaneous) laws of nature is limited to past descriptions of nature. That is, it allows determining that a description that was once accepted as true, whether it is a description of the past or a description of the present, was in fact false, and vice versa. Simultaneous laws of nature, on the other hand, allegedly allow doing the same, only regarding propositions about the present. However, this cannot explain the objectivity of the present, since any attempt to distinguish between what is currently held as true and what is actually true in the present results in a pragmatic contradiction. That is, it requires the conceptual scheme to acknowledge that a certain proposition is both currently held as true and is in fact false.

The objectivity of the present, therefore, can only be acknowledged when propositions about the present, that is, propositions of the type $N\varphi$, are considered from a later point of view, that is, in the form $P(N\varphi)$. To apply this distinction, between what was held as true and what was in fact true, to propositions of the type $P(N\varphi)$, all that is required is laws of nature that connect occurrences at different times. We can therefore only attempt to use simultaneous laws of nature to support the objectivity of propositions about the present when they are considered from a later point of view, in which case such laws of nature are superfluous. Hence, the only case in which such laws allegedly allow going beyond what non-simultaneous laws of nature allow, that is, to conclude what objectively exists in the present, leads to a pragmatic contradiction.

Again, it might be argued that this does not show that simultaneous laws of nature are not possible. Although such laws may not be necessary, because they are not necessary for explaining the objectivity of the present, they may nevertheless be possible. That is, although there may be limiting factors that do not allow drawing conclusions about what exists now based on simultaneous laws of nature, they may nevertheless play a role in formulating an objective description of nature in retrospect.

Although my main interest in this investigation is to explore the *necessary* features of nature, I believe exploring the suggestion that simultaneous laws may exist is important for understanding not only the structure of the laws of nature, but also their role in the constitution of nature. Suppose therefore that there are simultaneous laws of nature, which, although they cannot allow formulating inferential descriptions of the present at the present, can be used in retrospect to formulate inferential descriptions of nature. Thus, if such a law connects the simultaneous existence of φ and ψ, and at a certain time only φ is documented by the conceptual scheme, it is later possible to infer the existence of ψ.

Unfortunately, the suggestion to rely on simultaneous laws of nature to formulate, retrospectively, descriptions of nature is untenable. This is because such inferences can only be made in retrospect, that is, about the past, that is, if the conclusion is of the structure $P(N\psi)$. As such, as with all the propositions about the past, this description ultimately relies on non-simultaneous laws of nature. This dependence is reflected in two distinct ways.

To begin with, even if the subject matter of this inference is simultaneous occurrences, it ultimately relates to past occurrences. Hence, it relies on non-simultaneous laws of nature, which enable maintaining past descriptions in the present, that is, which constitute memory. Admittedly, this dependence is not necessarily problematic, given that this mechanism is supposed to reliably preserve descriptions in the conceptual scheme. Hence, there is no reason to suppose that it has any implication on the possibility of correctly identifying simultaneous laws of nature.

The more significant dependence of such an inference on the non-simultaneous laws of nature is that, because this inference relates to past occurrences, the conclusion of this inference is subject to the evaluation of the non-simultaneous laws of nature, as the criteria for objective descriptions of the past. Because the occurrences, which are described based on the simultaneous laws of nature, are in the past, these occurrences must be linked by the non-simultaneous laws of nature to current occurrences. Furthermore, these laws connect occurrences at different times, such that these occurrences are necessary and sufficient conditions for one another. Hence, the simultaneous laws of nature can play no role in describing nature. Only what is implied by the non-simultaneous laws of nature will be recognized as objectively existing. The laws of nature must hence be non-simultaneous, that is, they necessarily connect occurrences at different times.

5.13 Description of the Present

In this section, I wish to explore some of the implications of the conclusion of the previous section, which rejected the idea of simultaneous laws of nature, focusing mainly on the possibility of formulating descriptions of the present. Again, I wish to make clear that my discussion is conceptual. Hence, I shall not discuss problems such as the "specious present," and related issues which are associated with it (see, for example, (Arstila, 2018; Broad, 1923; James, 1890; Kelly, 2005; Le Poidevin, 2007). Furthermore, I believe that these problems arise from a misunderstanding of the conceptual structure that supports the distinctions between the past and the present, and hence because of confusion between conceptual and phenomenological questions.

Before beginning the inquiry, I wish to clarify that the conclusion of the previous section does not imply the impossibility of laws of nature that allow action-at-a-distance, such as Newton's law of gravity. According to this law, the effects of a gravitational force are instantaneous, regardless of the distance between the objects. Firstly, because space has yet to be introduced into the conceptual scheme I am developing here, the idea of action-at-a-distance has no meaning in this conceptual scheme. Furthermore, the term "instantaneous," as it is used to describe action-at-a-distance, is not equivalent to "simultaneous," as used in my description of "simultaneous laws of nature" in Sect. 5.12. As I have explained previously, in my use of this term, simultaneous laws are defined for a single point in time, and hence can apply even to a non-temporal realm. However, Newton's law of gravity determines force, and hence a change in time. It hence has no meaning in a non-temporal realm. The idea of instantaneous laws of nature will be discussed in Chap. 6, after the introduction of the concept of space.

To begin with, an implication of the impossibility of simultaneous laws of nature is that it is impossible to determine, based on occurrences at a certain time, and the laws of nature, about other occurrences at the same time. A consequence of this fact is that non-inferential descriptions of the present are limited to the subjective realm, that is, to the current state of the conceptual scheme.

To understand why, recall that my analysis of the ability to describe the past from a point of view in the present has uncovered the dependence of non-inferential descriptions of the past on the laws of nature, which connect occurrences at different times. These laws are the only laws available (as the previous section has shown). To allow for the unity of nature, these laws cannot be limited to a connection between one occurrence at each time. For this would create isolated chains of occurrences. Hence, I concluded that all occurrences must be connected by the laws of nature.

In light of this conclusion, I wish to examine the possibility of non-inferential descriptions of the present. I have argued that an objective conception of the present is a necessary feature of the conceptual scheme. That is, it is necessary to make a distinction between the truth-value that is ascribed to propositions in the conceptual scheme and what is actually the case. Admittedly, I have also argued that there is necessarily a subjective realm, which is the realm of the conceptual scheme itself, in which it is impossible to draw this distinction. However, what is under discussion here is the objective realm of the present.

It is easy to see now that a non-inferential description of the objective realm in the present is impossible. This is because the objectivity of this realm implies that its occurrences are distinguished from their simultaneous description in the conceptual scheme. As two distinct occurrences, the correspondence between the description and what is described requires an explanation, just as does the ability to describe the past from the present. Furthermore, the same explanation must apply to both cases. The laws of nature enable formulating objective descriptions of nature, that is, making the distinction between the description in the conceptual scheme and what is actually the case, and hence the laws of nature must also explain the connection between the two realms. These laws, however, are temporal, as they connect occurrences at different times. Hence, any non-inferential objective description of the present must be connected by these laws to what it describes, and hence actually determined by past occurrences.

I should emphasize the scope of this claim. Laws of nature are not simultaneous. That is, they do not connect simultaneous occurrences. Hence, any type of connection that is mediated by laws of nature must be non-simultaneous. The claim that non-inferential objective descriptions of the present are connected to past occurrences is only a particular instance of a general claim.

So far, I have discussed non-inferential descriptions of the present. The case with inferential descriptions of the present is different. Although inferential descriptions rely on laws of nature, just as do non-inferential descriptions, the dependence in the two cases is different. In inferential descriptions, the description is a conclusion of an argument, whose premises include the laws of nature and other descriptions of nature. These descriptions can be either inferential or non-inferential, but as previously explained, they all must ultimately rely on non-inferential descriptions.

Although all the non-inferential descriptions that are available at the present are descriptions of the past, it is in principle possible to rely on the laws of nature to infer from the non-inferential descriptions of the past what happens later, that is, in the present. However, the descriptions that are inferred in this way cannot be completely accurate. As will become evident in what follows, the conditions that determine the state of the conceptual scheme are necessarily different from the conditions that determine other occurrences in the present. Hence, the information that is available to the conceptual scheme cannot be used to accurately describe present occurrences.

I should emphasize that this does not imply skepticism about the ability to describe the present. It does imply that the objective descriptions of the present that are available to the conceptual scheme cannot be completely accurate. This gap between the conceptual scheme and what it describes will be found later to be of great importance for understanding nature.

Furthermore, I should point out that in this respect the description of the present is not fundamentally different from that of the past and the future, which are necessarily inaccurate. This is first and foremost due to the distinction between the subjective and the objective realms, that is, the distinction between the descriptions that are made by the conceptual scheme and nature. This implies that the description of nature is necessarily partial—as can be expected from the fact that the conceptual scheme is part of nature, and hence, as such, cannot reflect in its structure a complete and accurate description of all nature. The significance of this fact will be discussed further in Chap. 7.

What is unique about the description of the present is the fact that the non-inferential objective descriptions of what is alleged to be in the present are in fact descriptions of the past. In our conceptual scheme, these are identified with what is perceived by the senses. Although it is natural to believe that what we see is in the present, it follows from my analysis that what we see is necessarily in the past. However, I should emphasize that the claim I am making here is different from the claim the physicists make, according to which what we perceive now is a representation of what was the case in the past, due to the fact, for example, that the speed of light (or of sound) is finite. This is because I have not introduced yet the idea of space, nor did I discuss the idea of the speed of determination in nature. The conclusion that a non-inferential description can only be a description of the past follows directly from the idea of objectivity and the structure of the laws of nature. Furthermore, what scientists describe as a fact of nature is found to be a necessary feature of any conceptual scheme, which in what follows will be found fundamental for understanding nature.

The reader may wonder whether this implies a paradox in our conception of the present. Does this conclusion imply that the concept of the present is self-contradictory, by showing that descriptions of the present are really descriptions of the past? And if so, does this undermine the fundamental distinction between the past and the present?

The answer is negative. To begin with, there is a clear distinction between the past and the present, and there are available descriptions of the present, which not only accurately describe the present but are also infallible. These are the descriptions of the subjective realm in the present, that is, descriptions of the conceptual scheme itself. What I have argued is that non-inferential descriptions of the objective realm are necessarily descriptions of the past. Admittedly, it is a common prescientific belief that non-inferential objective descriptions of the present are not only possible but are in fact available to us. The fact that this belief is mistaken does not constitute a paradox.

A significant implication of this conclusion is that a perfect simultaneous correspondence between the conceptual scheme and nature is impossible. This is an implication of the distinction between propositions and what they describe, which was found to be necessary to explain the change in the distinction between past, present, and future, which themselves were found to be necessary to explain the need for contingent propositions for describing nature. It is therefore of the nature of existence that there cannot be a perfect match between different parts of it. I will return to this finding in what follows.

5.14 On the Impossibility of a Global Law of Nature

The laws of nature are contingent, and hence must be learned from experience. However, this fact alone does not explain the difficulty of identifying the laws of nature. Identifying the laws of nature is challenging because the non-inferential descriptions of nature do not enable identifying an unviolated regularity. This is necessarily the case, because if these descriptions would formulate an unviolated regularity, there would have been no need to make the distinction between the subjective and the objective realms.

The difficulty of identifying the laws of nature in experience is therefore implied by the possibilities of mistake, ignorance, and the non-existence of the conceptual scheme, which are necessary for explaining the need for contingent propositions for describing nature. It is thus easy to see the conceptual connections that mutually support all these ideas.

Nevertheless, as is evident from this chapter, it is possible to determine a priori the structure of the laws of nature, based on considerations that relate to the necessary conditions for describing nature with the help of contingent propositions. These considerations not only imply the distinction between the subjective and the objective realms, but also the need to identify the laws of nature, for it is necessary to rely on these laws to formulate inferential objective descriptions of nature.

When I first began analyzing the structure of the laws of nature, I relied on a minimalistic conception of these laws, which connects two properties at different times. According to such laws, φ exists at some time if and only if ψ existed before it. This was the minimally justified structure of the laws of nature at that stage of the inquiry, for at that point there was no reason to even assume that nature includes more than one realized property at any point in time.

However, the investigation into the nature of the laws of nature has shown that the conception of a law of nature that connects two isolated occurrences in different times is too simplistic. The laws of nature must be able to connect multiple occurrences at one time with multiple occurrences at another time. This is because my analysis had shown that it is necessary to incorporate into the conceptual structure that I am developing here, not only the idea of the simultaneous existence of multiple occurrences, but also the idea of all the occurrences in nature being interconnected by the laws of nature, that is, the idea of the unity of nature.

It is established that the laws of nature provide the connection that unifies nature, and that they can connect multiple simultaneous occurrences at each time with multiple other occurrences at another time. Hence, it seems that the most straightforward way in which laws of nature can connect all the occurrences in nature together is if all the occurrences at a certain time determine together all other occurrences in other times. Thus, it seems that all the requirements of the laws of nature can be fulfilled by one global law of nature.

A few explanatory notes are in order. To begin with I understand, by the term "a global law of nature," a law that necessarily connects all the occurrences at a certain time with all the occurrences at any other time, no matter how close they are in time. I stress this point since I have argued before that all the occurrences in nature must be connected by the laws of nature. However, the previous conclusion only implies that there cannot be a completely isolated chain of occurrences. A global law of nature implies, on the other hand, that all the occurrences at one time are connected to all the occurrences at any other time.

Furthermore, the idea of a global law of nature is that of a law that does not apply to a smaller-scale application. Since it determines anything that occurs at any time, it does not leave room for any smaller-scale laws, or indeed any other law of nature. That is, there is only one law of nature, which encompasses all the occurrences at any given time, and connects them to all the occurrences at any other time.

Notwithstanding its attractiveness, it can be shown that a global law of nature cannot support the formulation of an objective description of nature. This is first and foremost because, to rely on such a law of nature to extend the description of nature beyond what is available by non-inferential description, it is necessary to rely on a complete description of everything that exists at a certain time. However, the objectivity of nature implies that even if it is possible to acquire a description of everything that exists at a certain time, it is impossible to determine that this is indeed a complete description unless this conclusion relies on the laws of nature, which only begs the question. In other words, it is possible to rely on a global law of nature only if complete knowledge of nature is possible, but if this knowledge is possible, laws of nature are useless for extending the description of nature.

It can therefore be concluded that the laws of nature, which connect occurrences at different times, must be restricted, and only connect some occurrences at different times. This is the only structure of the laws of nature that can be both identified and used for formulating objective descriptions of nature.

The conclusion that a global law of nature is impossible requires further elaboration and explanation. To begin with, there must be a way of determining which occurrences at each time are relevant for determining one another. This will be explored in Chap. 6, in which the idea of space will be developed. Nevertheless, it might be thought that the conclusion of this section has implications regarding our conceptual scheme, which does include the concept of space. More specifically, it might be thought that it is possible to infer from the analysis of this section that action-at-a-distance laws of nature are impossible. Seemingly, if all things are tied together by the laws of nature, and the connection is not limited by distance, it might be thought that everything that exists at one moment in time must be relevant for determining any previous or late occurrence—which was shown to be impossible. Hence, one may falsely determine that action-at-a-distance violates the conclusion of the previous analysis.

However, this line of reasoning is flawed. All that follows from my analysis in this section is that not all simultaneous things can be relevant for determining occurrences at other times. This does not preclude the possibility that only some occurrences, including non-local, would be relevant for determining occurrences at other times. Only once the idea of space is introduced, and the notion of distance in space is explained, would it be possible to reject the idea of action-at-a-distance.

To conclude this section, I wish to summarize some of the implications of my analysis, as well as clarify some of the limiting conditions on the criteria for determining which occurrences are tied together by the laws of nature. Recall that I have argued that the laws of nature connect occurrences at different times, such that these occurrences are necessary and sufficient conditions for one another. Furthermore, the laws of nature are responsible for the unity of nature, and hence must connect all occurrences, although not necessarily directly. More specifically, they must be able to connect multiple occurrences at any given time to occurrences at other times. However, the previous analysis shows that not all the occurrences at a specific time can be relevant for determining occurrences at other times.

Suppose that occurrences at a certain time *t'* determine together later occurrences at a later time *t"*. A criterion is hence necessary for determining which occurrences at *t'* are relevant for determining which occurrences at time *t"*. Notice that although not all the occurrences at time t' can be relevant for determining any occurrence at time t", all occurrences at time t' must take part, in some form or another, in determining some occurrences at time t". This is because an occurrence φ at time t' that does not take part in determining later occurrences will not be identifiable in the future. That is, the occurrences at time t" together with the laws of nature would not be able to objectively determine its existence.

I have argued that the laws of nature must be able to connect different occurrences in the determination of future occurrences. This does not imply that a single occurrence at time *t'* cannot determine a single occurrence at time *t"*. The criterion for determining which occurrences at time t' determine which occurrences at time *t"* can allow for connections between single occurrences at different times, as long as it also allows for the possibility of connecting multiple occurrences at different times in some circumstances.

Furthermore, there can be true generalizations that are not formulated in terms of necessary and sufficient conditions between occurrences at different times and that are not accidental. An example of such a true generalization is "All uranium spheres are less than a mile in diameter" (Van Fraassen, 1989, p. 27). These true generalizations do not express laws of nature, but if they are indeed not accidental, they must rely on laws of nature, which do connect occurrences at different times.

What is important to keep in mind is that the connections between occurrences at different times must be such that the past and the future are both necessary and sufficient conditions for one another. That is, the laws of nature and the relevant criterion for the connections between occurrences at different times must be such that it is possible in principle to identify all past occurrences based on future occurrences. For, as I have argued before, existence is ultimately determined objectively according to the laws of nature.

The task of providing the criterion for determining which occurrences at a certain time are linked to which occurrences at other times is undertaken in Chap. 6. The answer to this question will be found with the introduction of space into the conceptual structure that is developed here.

5.15 Contingency and Necessity in the Laws of Nature

I have argued that the laws of nature are contingent, and hence must be based on non-inferential descriptions of nature. Nevertheless, I have also argued that the existence of these laws is a necessary feature of nature, and I have further described their necessary structure. This dual aspect of the laws of nature may raise a justified suspicion and a rightful expectation for an explanation, as to how these two seemingly contradictory features of the laws of nature can cohere.

To begin with, the seeming contradiction can be easily solved by introducing a familiar distinction between form and content. Thus, while the existence of the laws of nature in general, as well as their form, is a necessary feature of nature, the specific laws of nature, due to their contingent content, are themselves contingent.

While this answer is valid, it leaves unaddressed the tension between contingency and necessity, which accompanies the current investigation. Recall that my analysis is intended to explain the need for contingent propositions for describing nature. This investigation, however, is a conceptual investigation, which purports to expose the necessary structure of nature. I began my inquiry in Chap. 4 by uncovering the structure of time as a necessary feature of nature, while in this chapter I outlined the necessary structure of the laws of nature.

For my analysis, I have relied on the assumption that contingent properties, or occurrences, whose nature was not explained, do exist in nature. I have developed the conceptual structure that is required for describing the change in these properties, but I have not explained how the two aspects of our descriptions of nature cohere. Not only does the origin of the contingent properties remain unexplained, but it might also be argued that it is unintelligible that contingent properties exist side by side and in perfect match with the necessary features of nature, as is reflected in the structure of time and the laws of nature. Furthermore, if there are necessary features of nature, should they not characterize any element of nature, including the so-called "contingent" properties? The laws of nature therefore only highlight a general problem, because they incorporate these seemingly inconsistent features of nature.

In this context, it is important to remember another tension, if not outright contradiction, that arose in the course of my analysis. This is the tension between the discovery that existence implies a description from a point of view of a conceptual scheme, and hence existence is relative, and the fact that these descriptions imply a distinction between themselves and what they describe, and hence imply an objective conception of nature, such that it is independent of the conceptual scheme from whose point of view it is described.

The reason I am addressing this tension here is that it is fundamentally connected to the tension between contingency and necessity. In fact, the solution to the tension between relativism and objectivity is also the solution to the tension between contingency and necessity.

To begin with, the fact that the conceptual construct that was introduced in the course of this inquiry is a necessary structure of existence implies that it is relevant for any existing entity, whether it is nature as a whole or individual occurrences or properties in nature. However, it does not follow that it is impossible to find room for contingency in this conception of existence. As soon becomes evident, it is the distinction between nature and the point of view from which it is described, which necessarily provides a limited point of view of nature, that explained the contingency of propositions in language and of the contingent aspect of laws of nature in particular. Contingency reflects the limited point of view of nature.

An example might be helpful here. In the history of philosophy, one of the most ancient debates revolved around the question of whether matter is composed of indivisible components, that is, atoms, or not. Suppose that this question is answerable by conceptual considerations and that a conclusive argument is found, which establishes univocally the existence of atoms. Hence, the claim that matter is made of atoms is a necessary truth. However, this conclusion cannot identify atoms in nature, that is, the actual entities that are indivisible. Indeed, the physical entities which are now identified as "atoms" are so named because they were mistakenly thought to be indivisible. Thus, even if it is known that atoms necessarily exist, the claim that *this* entity in nature is an atom, that is, indivisible, is a contingent proposition. Thus, the contingency of propositions in language can be explained by the fact that these propositions reflect the conceptual scheme's partial and limited point of view of nature.

As the example of atoms shows, the distinction between the description of nature and what is actually the case refers not only to the distinction between a partial and a comprehensive description of nature. It also applies to the distinction between different levels of descriptions. That is, the identification of the fundamental level of nature is of a contingent nature. Indeed, my analysis of the non-inferential descriptions of nature has shown that ascriptions of truth-values are not basic properties of nature, and hence cannot feature in the fundamental laws of nature (although they must ultimately rely on the laws of nature). This demonstrates how nature can be described on different levels, and at some levels of description, the laws of nature need not be represented directly. Again, the level of description is determined from the point of view of the conceptual scheme, whose own descriptions do not reflect the most fundamental level of nature. The properties that are described by non-inferential descriptions of the conceptual scheme are hence expected to be emergent properties, that is, higher-level properties of nature. Therefore, any identification of these properties must likewise be formulated by contingent propositions.

Furthermore, my analysis has shown that nature is necessarily subject to the laws of nature. However, from the limited point of view of the conceptual scheme over nature, the laws of nature cannot be determined conclusively. On the contrary, the identification of the laws of nature must rely on non-inferential descriptions of nature, which offer a limited description of nature. Moreover, it is the assumption that nature is subject to the laws of nature that allows for formulating an objective description of nature. Hence, any identified law of nature is necessarily contingent.

It can be concluded that the fact that the specific laws of nature are described by contingent propositions, although the laws of nature reflect a necessary feature of nature, is an implication of the limited point of view of the conceptual scheme over nature. I should emphasize that this feature of the laws of nature is not something avoidable. The contingent aspect of the laws of nature is a necessary feature of nature. It reflects the fact that existence is relative, and determined in relation to a point of view that is internal to nature. This conception of nature will be further explored and explained in Chap. 7.

The claim that the contingency of propositions in a language is explained by the point of view from which nature is described is prima facie similar to Leibniz's explanation of contingency of propositions in language, as a result of the limitations of our finite minds (Leibniz, 1989, p. 28). This is contrasted with the mind of God, for whom all truths are necessary. However, I should note that there is a fundamental difference between the two views. First and foremost, according to my analysis of meaning, contingent propositions form the foundation of meaning in language and do not reflect any limitation of our minds. It is explained by the fact that nature is described from a point of view that is internal to it, which is reflected in the use of indexical expressions, which are irreducible, as I have argued in Sect. 4.10.

Finally, I have argued that the end of human knowledge will be achieved when the metaphysical a priori understanding and the physical empirical knowledge of nature will be brought together and unified. As should by now be clear, both research projects are indispensable for formulating a comprehensive conception of nature. However, the final identification of the necessary features of nature, as uncovered by metaphysics, with the contingent properties of nature, as discovered by physics, will itself be formulated by contingent propositions.

References

Armstrong, D. M. (1978). *A theory of universals*. Cambridge University Press.
Armstrong, D. M. (1983). *What is a law of nature?* Cambridge University Press.
Armstrong, D. M. (1991). What makes induction rational? *Dialogue, 30*(4), 503–511.

Armstrong, D. M. (1993). The identification problem and the inference problem. *Philosophy and Phenomenological Research, 53*(2), 421–422.
Arstila, V. (2018). Temporal experiences without the specious present. *Australasian Journal of Philosophy, 96*(2), 287–302.
Augustine. (2019). *Confessions* (T. Williams, Trans.). Hackett Publishing Company, Incorporated.
Bird, A. (2005). The dispositionalist conception of laws. *Foundations of Science, 10*(4), 353–370.
Broad, C. D. (1923). *Scientific thought*. Routledge and Kegan Paul.
Cartwright, N. (1980). Do the laws of physics state the facts? *Pacific Philosophical Quarterly, 61*(1/2), 75–84.
Dretske, F. I. (1977). Laws of nature. *Philosophy of Science, 44*(2), 248–268.
Earman, J. (1984). Laws of nature: the empiricist challenge. In R. J. Bogdan (Ed.), *D.M. Armstrong* (pp. 191–223). Springer.
Fales, E. (1990). *Causation and universals*. Routledge.
Friedman, W. J. (1990). *About time: inventing the fourth dimension*. MIT Press.
Giere, R. N. (1999). *Science without laws*. University of Chicago Press.
Goodman, N. (1983). *Fact, fiction, & forecast* (4th ed.). Harvard University Press.
Hume, D. (1978). In L. A. Selby-Bigge (Ed.), *A treatise of human nature*. Oxford University Press.
Jackson, F., & Pargetter, R. (1980). Confirmation and the nomological. *Canadian Journal of Philosophy, 10*(3), 415–428.
James, W. (1890). *The principles of psychology*. Henry Holt.
Kant, I. (1929). *Critique of pure reason* (N. Kemp-Smith, Trans.). Macmillan.
Kelly, S. D. (2005). The puzzle of temporal experience. In A. Brook (Ed.), *Cognition and the brain: The philosophy and neuroscience movement* (pp. 208–238). Cambridge University Press.
Lange, M. (1993). Natural laws and the problem of provisos. *Erkenntnis, 38*(2), 233–248.
Lange, M. (2000). *Natural laws in scientific practice* (Vol. 71). Oxford University Press.
Le Poidevin, R. (2007). *The images of time: An essay on temporal representation*. Oxford University Press.
Leibniz, G. W. F. (1989). In R. Ariew & D. Garber (Eds.), *Philosophical essays*. Hackett Publishing Company.
Lewis, D. K. (1973). *Counterfactuals*. Blackwell.
Lewis, D. K. (1983). New work for a theory of universals. *Australasian Journal of Philosophy, 61*(4), 343–377.
Lewis, D. K. (1986). *Philosophical papers* (Vol. 2). Oxford University Press.
Lewis, D. K. (1994). Humean supervenience debugged. *Mind, 103*(412), 473–490.
Loewer, B. (1996). Humean supervenience. *Philosophical Topics, 24*(1), 101–127.
Mumford, S. (2004). *Laws in nature*. Routledge.
Poppel, E. (1978). Time perception. In R. Held (Ed.), *Handbook of sensory physiology* (Vol. VIII). Springer.
Russell, B. (1921). *Analysis of mind*. Routledge.
Shoemaker, S. (1980). Causality and properties. In P. Van Inwagen (Ed.), *Time and cause* (pp. 109–135). D. Reidel.
Shoemaker, S. (1998). Causal and metaphysical necessity. *Pacific Philosophical Quarterly, 79*(1), 59–77.
Swoyer, C. (1982). The nature of natural laws. *Australasian Journal of Philosophy, 60*(3), 203–223.
Tooley, M. (1977). The nature of laws. *Canadian Journal of Philosophy, 7*(4), 667–698.
Tooley, M. (1987). *Causation: A realist approach*. Oxford University Press.
Van Fraassen, B. C. (1989). *Laws and symmetry*. Oxford University Press.
Ward, B. (2002). Humeanism without humean supervenience: A projectivist account of laws and possibilities. *Philosophical Studies, 107*(3), 191–218.
Yehezkel, G. (2013). The Illusion of the experience of the passage of time. *Disputatio, 5*(35), 67–80.
Yehezkel, G. (2014). *The conceptual structure of reality*. Springer.

Chapter 6
The Structure of Time and Space

Abstract In this chapter, I define the structure of time and space. I begin by constituting time as an objective feature of nature, by showing how, based on the laws of nature, the relations of simultaneity and precedence can be given objective meaning, and how distances in time can be measured. Next, I introduce and develop the notion of space by giving meaning to the idea of locality and defining spatial relations in terms of the temporal order of reciprocal determinations. Finally, I show that there is a fundamental constant speed of determination between occurrences in space, and further explain how this can be used for measuring space. The entities in space and time are shown to be reciprocally determining elements, which are themselves constituted by temporo-spatial relations.

Keywords Laws of nature · Locality · Measurement · Space · Speed · Time

6.1 Introduction

In the course of the previous two chapters I have uncovered the necessary structure of existence, which is constituted by the change in the temporal distinctions of past, present, and future, and in the laws of nature, which connect occurrences at different times. As a necessary structure, it must ultimately constitute all the properties of nature. This conclusion has implications for the structure of the fundamental laws of nature. So far, my discussion has relied on a distinction between form and content, that is, between the structure of the laws of nature and the properties which are connected by these laws. This was an inevitable implication of the attempt to reconstruct the conceptual apparatus which allows the formulation of contingent propositions. However, by now it should be clear that this distinction is artificial, and cannot apply to the fundamental laws of nature. The properties in nature cannot be separated from the necessary conceptual structure which supports the concept of existence.

Nature is hence constituted by a temporal structure, which allows a change in properties, which are themselves constituted by temporal relations. This is a circular

G. Yehezkel, *Conceptual Principles of Natural Philosophy*, Synthese Library 495, https://doi.org/10.1007/978-3-031-82108-0_6

characterization. However, this circularity is an expected implication of the fact that this is a characterization of a necessary structure, and hence can only be described by a tautology. Nevertheless, this circular description of the fundamental structure of time is unsatisfactory because it is trivial, as, for example, is the claim that all bachelors are bachelors. The circularity is so narrow that it has very little explanatory value. To be of value, and offer an insight into the nature of existence, it is necessary to formulate a more detailed account of this structure, which would allow opening a gap between time and the basic properties of nature, and thus allow giving content to the idea of properties which are constituted by temporal relations.

For this purpose, it is necessary to introduce at least one new (necessary) property that, although ultimately constituted by time, is defined in a way that allows it to change in time. A clue to the identity of this property can be found in the idea that nature must be described from a point of view, that is, a conceptual scheme, that is internal to nature. This implies that existence is relative, and depends on relations that obtain between the entities that coexist in nature. Coupled with the conclusion of the previous chapter, according to which all things in nature must be tied together by the laws of nature, it follows that a change in time is constituted by the reciprocal determination of the coexisting entities. In the course of my inquiry, this tentative suggestion will be found to be correct, and this feature of nature, which gives substance to the idea of a change in time, will be identified with space.

Before beginning the analysis, I should again stress that the investigation I am undertaking here is conceptual, and hence should not be confused with phenomenological research of temporal experience, or the causal mechanisms which are associated with it (see, for example, Dennett, 1991; James, 1890; Mellor, 1998). Notwithstanding the merits of empirical research of the experience of time, it has no place in conceptual analysis. I should acknowledge, on the other hand, that my line of inquiry in this chapter is greatly inspired by Strawson's analyses in his *Individuals* (Strawson, 1959). Strawson's arguments are themselves inspired by Kant's arguments in the "Analogies of Experience" and in the "refutation of idealism" (Kant, 1929, pp. B218–B265, B274–B279). Unfortunately, their study of time and space, and the function that they fulfill on the constitution of an objective view of nature, cannot be integrated into the current analysis, which is constituted on the conceptual structure that was developed in the previous chapters, and formulated in terms of the conception of language as a system of coordinates.

6.2 Defining Simultaneity and Precedence

To articulate an objective view of nature, it is necessary to formulate descriptions that do not include the indexical expression of past, present, and future. Even though these distinctions are the basic distinctions that explain the meaningfulness of propositions in language, they are subjective in the sense that they reflect the fact that nature is described relative to a location in time that is internal to it, that is, the point of view of a conceptual scheme. Furthermore, these distinctions per se are not accurate enough to allow describing positions and distances in time, because only the temporal indexical expression "present" refers to a point in time. The indexical terms "past" and "future" are general terms that apply to a multitude of occurrences that occur at different times. Hence, the most important task is to introduce terms that allow making more detailed distinctions than those of past, present, and future. This would enable formulating accurate descriptions of locations and thus of distances in time.

In Sect. 4.13, I explained how the distinction between present and past can create a series of ordered occurrences in time. This series organizes the occurrences in time by their order. This is the basis for the distinctions of simultaneity and precedence, which are the temporal relations that obtain between all the occurrences in nature. These relations make it possible to create an objective conception of nature, because they allow the formulation of descriptions that are unaffected by the temporal location of the conceptual scheme.

The first step in formulating an objective description of nature hence consists in defining the relations of simultaneity and precedence, based on the distinction between the past and the present. Although this does not allow objective determination of the notions of precedence and simultaneity, because this definition relies on the distinctions between past and present, which are determined in relation to the point of view of a conceptual scheme, it is the first step in giving meaning to an objective determination of these terms.

I should emphasize that the current investigation cannot rely on a conceptual structure that was not yet introduced. Hence, I rely only on the conceptual apparatus that was introduced in Chap. 4. Thus, ψ precedes φ if and only if $(N(N\varphi \wedge P\psi) \vee P(N\varphi \wedge P\psi) \vee F(N\varphi \wedge P\psi))$, that is, ψ precedes φ if and only if it now, was, or will be the case that φ is present and ψ is past.

Notice that this definition does not relate to what was accepted as true in a conceptual scheme. Rather, it relates to the actual objective state of the world. Hence, this definition is suitable for an objective description of nature. Further, notice that the notion of precedence that is defined here allows formulating descriptions of nature in another objective sense, that is, it allows formulating propositions whose truth-value does not change with the context of their temporal use, and hence they do not reflect the point of view of a conceptual scheme. In contrast to propositions that include temporal indexical expression, for example, $N\varphi$, the truth-value of the proposition "ψ precedes φ" is unaffected by the time of its use.

Similar considerations enable defining the relation of simultaneity. Because the only temporal indexical expression that is not general is "present," it is the only term that can uniquely define this relation. According to this definition, ψ is simultaneous with φ if and only if $(N(N\varphi \wedge N\psi) \vee P(N\varphi \wedge N\psi) \vee F(N\varphi \wedge N\psi))$, that is, ψ is simultaneous with φ if and only if it now is, was, or will be the case that both ψ and φ are present. Again, notice that this definition is objective both in the sense that it does not refer to what was accepted as true in a conceptual scheme, and in the sense that it allows formulating propositions that are not sensitive to the context of their use.

6.3 Objective Determinations of Simultaneity and Precedence

So far, I have referred to the conceptual aspect of defining the relations of precedence and simultaneity. This provides the necessary concepts for formulating time in the form of an objective series of occurrences. It is now time to ask, how can the existence of these relations be determined objectively? Notice that I do not ask how temporal order is observed (see, for example, Dennett, 1991; Le Poidevin, 2004; Mellor, 1998). Rather, I ask what constitutes objective temporal relations.

Even though it seems possible to directly observe the temporal relations that obtain in nature, the determination of objective temporal relations cannot rely on the non-inferential description of nature alone. This is because these descriptions reflect the point of view of the conceptual scheme. However, even regarding the present, the possibilities of mistake and ignorance imply that the non-inferential descriptions are insufficient for formulating objective descriptions of nature.

To determine the existence of objective relations, it is necessary to rely on the laws of nature, which constitute the objective realm. Beginning with the relation of simultaneity, it is important to keep in mind that the possibility of simultaneous laws of nature was previously discussed and rejected (in Sect. 5.12). The laws of nature must connect occurrences at different times, such that the occurrences in these times are necessary and sufficient conditions for one another.

However, it was also shown that all the occurrences in nature must be connected (in Sect. 5.11). This is a necessary condition for the unity of nature, which does not allow acknowledging the existence of what is not connected by the laws of nature to other occurrences in time. The laws of nature cannot, therefore, be limited to only individual occurrences at each time. On the contrary, the laws of nature must be able to connect multiple occurrences at one time with multiple other occurrences at another time, such that multiple occurrences take part in the determination of past or future occurrences. Hence, it is possible to determine objectively that two occurrences are simultaneous if and only if these occurrences determine together, via a law of nature, a later occurrence that immediately follows them.

Notice that the last restriction, that is, "that immediately follows them," is necessary because I have argued for the unity of nature, that is, that the laws of nature connect all occurrences in nature. Two occurrences at different times can therefore together take part in the determination of a single later occurrence. Hence, this characterization depends on the possibility of giving meaning to the determination of distances in time.

Furthermore, recall that I have shown that the idea of global laws of nature is untenable, that is, there are no laws of nature that incorporate all the occurrences at a certain time in determining a later occurrence. That is, the laws of nature must be limited in their scope, that is, "local" in spatial terms. However, this does not imply that it is impossible to determine global simultaneity. This is because simultaneity is a transitive relation, and hence can be extended by continually connecting adjacent occurrences, thus covering occurrences at distant parts of nature. This is the reason why my characterization of simultaneity does not include the restriction "only if." Two occurrences can be simultaneous even if they do not determine together, via a law of nature, any immediately following occurrence. This is possible if they are connected by a chain of occurrences, in which every two adjacent occurrences determine together an immediately following occurrence. Thus, an occurrence ψ is simultaneous with occurrence φ if they are both members of an ordered series of occurrences, such that every two adjacent occurrences are locally simultaneous with one another. Admittedly this characterization depends on the ability to give meaning to the idea of locality, and hence also to that of space, as an order of coexisting occurrences. This task will be undertaken in what follows.

Another limitation of this characterization is that it depends on the existence of a series of occurrences that connect two remote occurrences. However, there may be simultaneous occurrences that cannot be linked in this way, due to the lack of intermediate occurrences. However, once other notions are introduced—the idea of measuring distances in time, the idea of space, and finally, the idea of the speed of determination—it would be possible to extend this characterization to include occurrences that are not local, according to the way that they take part in the determination of occurrences, which do not necessarily follow them immediately. These ideas, however, necessarily rely on the characterization of simultaneity which I introduce here.

Although my investigation in this book is purely conceptual, and discussion of physical theories extends beyond the scope of this discussion, a cautionary remark is called for in this context. I should make clear that this characterization of global simultaneity does not contradict Einstein's Special Theory of Relativity, because it is, as yet, unclear what its implications are, specifically whether it implies that this relation is the same for all observers. Although a conceptual investigation should be undertaken in complete disregard of the empirical investigation of nature—a necessary condition for its integrity—and should not shy away from contradicting contemporary accepted physical theories, stressing this point is important as a clarifying note, regarding the meaning of this characterization.

Finally, it should be noted that the characterization of the objective relation of simultaneity is also a straightforward characterization of distance 0 in time, that is, of being in the same location in time. Although this may sound trivial, it nevertheless demonstrates how time measurement can be given meaning by the laws of nature.

Turning the attention to the relation of precedence, the objective determination of precedence can be characterized straightforwardly based on the laws of nature. Two occurrences ψ and φ precede one another if there is a law of nature that connects these two occurrences, such that ψ and φ are, alone or together with other occurrences, necessary (or sufficient) conditions for one another.

Notice that this characterization does not determine which occurrence is earlier and which is later. This reflects the fact that the laws of nature are symmetrical, and do not distinguish between earlier and later. It is possible to rely on the definition of precedence, which was introduced in the previous section, to determine which occurrence is earlier and which is later. However, this relies on a relation to a point of view of a conceptual scheme, and hence is subjective. To determine objectively which occurrence precedes and which follows, it is necessary to rely on an asymmetry in the way in which the occurrences at different times determine one another. The asymmetry between the past and the future will be explored in Sect. 7.5.

Again, this characterizes precedence only locally. However, just as in the case of simultaneity, it is possible to extend this characterization. By relying on the notion of global simultaneity it is possible to extend the relation of precedence to occurrences that are not connected directly by a law of nature. Thus, two occurrences ψ and φ precede one another if and only if ψ is simultaneous with another occurrence μ, and there is a law of nature that connects μ and φ, such that μ and φ are, alone or together with other occurrences, necessary (or sufficient) conditions for one another.

6.4 The Use of Cyclical Regularities for the Measurement of Time

The measurement of distances in time requires the ability to determine locations in time, as well as a measuring unit. Both the locations and the measuring unit must be able to be determined objectively. Objectivity is constituted by the laws of nature. Hence the laws of nature must determine both locations and the measuring unit.

It is not coincidental that the laws of nature enable identifying locations and provide a measuring unit, thus allowing determination of distances in time. The ability to determine distances in time is necessary for formulating objective temporal determination. Firstly, recall that even the determination of objective simultaneity relies on the notion of an occurrence that *immediately* follows another occurrence. Furthermore, relying not only on the laws of nature to make objective distinctions requires a way of determining distances in time. Even identifying the laws of nature requires the determination of distances in time, for the laws of nature are supposed to connect *adjacent* occurrences in time. Otherwise, there is no way of determining which occurrences are connected by the laws of nature. Hence, the identification of the laws of nature should rely on the ability to objectively determine that two occurrences are adjacent in time. Without giving meaning to the objective determination of proximity in time, it seems always possible that more occurrences exist between two observed successive occurrences. Thus, for example, if φ is observed to follow ψ, it is always possible that another occurrence μ, which was not recorded, existed between them. Without the ability to measure time, it is impossible to rule out this possibility.

However, it might be thought that this example is too simplistic and can be further developed. Thus, for example, even David Hume, a devout empiricist, has accepted the possibility of imagining a shade of a color that was never experienced, based on imagining a gradual change in shades of this color that was experienced (Hume, 1978, p. 6). Similarly, it seems possible to infer, after noticing light blue followed by a dark blue, that there was a missed point in time in which the color was blue. This offers a way to solve the problem of the existence of time lapses between two identified properties.

However, once it is accepted that it is possible that some of the members of this series have not been noticed, it is impossible to determine how many shades of color have been missed. The possibility of ignorance opens the possibility of an unlimited number of members in that series. Furthermore, it is possible that many changes occurred between the two shades of blue that were observed. For example, the color may have changed from dark blue to light blue several times, time may have elapsed without any change in the color, and so on. There is no way to determine what happened between two seemingly adjacent observations.

Notice that the stipulation that the actual change in shades of blue is constant cannot be justified. For even if a constant change in the shade of color is observed, nothing negates the possibility that there were changes that were missed, in the course of what seems to be a constant change. Furthermore, the same problem arises if one resorts to the idea of multiple coexisting series of occurrences, such that one regularity can function as a measuring tool, in relation to which other series of occurrences can be measured. For if the problem of determining objective occurrences in time is not solved by one series, resorting to multiple series would not solve the problem either. The same difficulties that arose regarding the first series would be relevant to any other series as well.

It may seem that the problem I am raising here leads to total skepticism. After all, due to the unique function of the temporal distinctions between past and present, on which all the temporal distinctions rely, it is impossible to directly identify the temporal determinations. Rather, they must rest on the identification of properties that are realized in time. If the identification of properties in time is insufficient for determining distances in time, it may seem that determining the distances in time is impossible.

In response, it is necessary to make a distinction between the epistemological question and the metaphysical question. This distinction can be demonstrated by the problem of the uniformity of nature, which was discussed in Sect. 5.8. The problem of the uniformity of nature is sometimes misunderstood, and Hume is blamed for confusing induction with deduction, or for asking for something impossible, that is, a way to assure that our future predictions will be found to be true (Goodman, 1983, p. 62). However, this is a wrong interpretation of Hume's critique of induction. Hume does not seek an assurance that what we identify as a law of nature is indeed a law of nature. As I explained, such identifications are contingent, and hence always susceptible to the possibility of mistake. What Hume asks is to justify the assumption that underlies all such contingent identifications, and this is the assumption that nature is subject to laws of nature at all.

Similarly, it is obvious that because time is not perceived per se, due to the unique function of the temporal distinctions, all attempts to determine objective temporal distinctions must rely on observed regularities. Hence, any such claim must be contingent, and susceptible to the possibility of mistake. However, the problem is to establish a connection between some properties in nature and the determination of distances in time. The problem is therefore not that of identification, but rather of the structure of the laws of nature, and its connection to the constitution of objective temporal determinations.

The difficulty to which I am turning attention is raised because the properties that I have described so far, including colors, are present in a way that is divorced from fundamental temporal distinctions. These properties cannot be fundamental. Rather, they must be identified as high-level properties, that is, as complex properties that ultimately rely on the fundamental temporal structure that was found to be a necessary feature of anything that exists. I have so far stipulated the existence of properties whose nature was not explained, and relied on a simplistic view of the laws of nature as a successive series of such properties, which are seemingly independent of any temporal distinctions. This is clearer in the case of colors that characterize our phenomenological space, which seem atomic and therefore unanalyzable; hence their connection with temporal distinctions is yet to be uncovered.

As already mentioned, all properties in nature must ultimately rest on the temporal apparatus which was found to be a necessary feature of nature. Hence, the laws of nature must be formulated in terms of properties that are constituted by a change in time. This will allow me to offer a solution to the skeptical challenge I have presented above. That is, once the properties which change in time are identified, and explained to be constituted by the change in time, there is no room to doubt how much time has passed. These properties will be identified in the following sections with spatial properties.

As a final note on this subject, I shall note that the fact that the laws of nature constitute time (and space) explains why these laws are uniform and applicable to any part of nature. Thus, although the laws of nature can be uncovered by empirically studying a limited section, in time and space, they apply to nature as a whole, thus allowing the expansion of the description of nature beyond what is given by non-inferential descriptions, that is, beyond what is directly observed in experience.

At this stage, I believe that the need for determining objective distances in time is clear. For this purpose, it is necessary to resort again to the laws of nature. Since the laws of nature were introduced to enable the formulation of an objective description of nature, it is not surprising that they also provide the conceptual resources that are needed for determining distances in time. Indeed, the solution for determining distances in time is given by the very nature of the laws of nature. Recall that the laws of nature are supposed to provide the means for extending the description of nature beyond what is immediately given, based on connections that obtain between occurrences at different times. For this purpose, for example, in the case that occurrence of type φ is observed, to determine that occurrence of type ψ preceded it, it is necessary to rely on a regularity that obtains in nature, which ties these occurrences in a lawful connection. As I have argued, because these laws are contingent, and must be learned from experience, they must describe recurring regularities. Thus, after witnessing a pattern in which ψ precedes φ, it is possible to infer that an unobserved occurrence of type ψ occurred before an observed occurrence of type φ. It is therefore a necessary feature of the laws of nature that they describe recurring regularities in nature. Since the laws of nature constitute the objective realm, these regularities constitute objective temporal determinations.

To determine distances in time, it is necessary to rely on a measuring unit. To be useful, this measuring unit must not only be a recurring regularity, but also a cyclical regularity, which can continually repeat for indefinitely long periods of time. For practical propose, it suffices to rely on a higher order regularity, such as the position of the sun in the sky, which although it ultimately relies on the laws of nature, does not guarantee perfect accuracy. That is, these repeating cycles are not necessarily identical in length. However, a fundamental accurate regularity, which directly reflects the laws of nature, must exist in nature and is accurate by definition—all the cycles that are constituted by such regularity are necessarily identical in length since they constitute time itself.

The cyclical regularities that are used as a measuring unit allow the measurement of overlapping processes by mutual comparison. In this comparison, it is necessary to rely on the laws of nature in another respect: that is, to determine relations of simultaneity between occurrences in the cyclical regularity and the overlapping measured process. Thus, distance in time can be determined.

Notice that the determination of distances in time, in comparison to recurring cyclical regularity, allows measuring lengths of time as a relation to the unit of measurement. Hence, the fact that a certain cycle is used as a unit does not imply that the measurement of time is discrete. Theoretically, there is no limit to the accuracy of measurement of this type, and hence this way of measuring time is compatible with a continuous perception of time. Although I shall not pursue this intricate subject here, I note in passing that a continuous view of time seems to cohere with the conception of existence which emerges from my analysis, which identifies existence, as well as the most fundamental features of existence, with relational properties.

Before concluding this analysis, I wish to clarify a possible misunderstanding. My analysis implies that there must be a law of nature that determines that a certain fundamental change in nature is always constant. To avoid any confusion, I should emphasize that at this stage of the inquiry it is impossible to identify this process with any particular change in nature, and more specifically with a spatial change, that is, speed. All it implies is that there is a constant change in nature, which constitutes a law of nature. The question of speed can only be addressed after the introduction of the notion of space.

Finally, I would like to remark on the significance of the conclusion of this section, which might seem trivial. To begin with, it may be claimed that it is not surprising that the laws of nature allow determining distances in time, for this is necessary for an objective description of nature, and this is exactly what the laws of nature are supposed to provide. Furthermore, the claim of this section does not seem innovative, given the fact that the human race has been measuring time based on cyclical regularities for millennia.

To clarify, the significance of the conclusion of this section is in establishing not only that there must be a way to measure time, but also in showing that the conceptual structure that was introduced so far is sufficient for supporting this possibility. Furthermore, my analysis shows that there must be a regular change in nature that provides an accurate measurement of time, which has the status of a law of nature.

The accuracy of this yet-to-be-determined change is guaranteed by the fact that the law of nature that determines this change constitutes time itself.

6.5 Relations of Determination Between Existents

I have argued that the laws of nature must connect occurrences at different times. However, I have yet to characterize the fundamental properties that feature in these laws. Nevertheless, it is clear that, as with any existent, these properties must be constituted by time, which itself is founded on the change in the fundamental properties of nature. In other words, these are not only the properties that change in time, but the change in these properties is what constitutes time.

This characterization of the relations between time and the fundamental properties that change in time is admittedly circular. However, as I have previously explained, the difficulty is not due to the circular nature of this connection, but because this circularity is too narrow. The laws of nature cannot be formulated merely in terms of changing temporal properties. In this section, I begin exploring these properties.

I have shown that an objective conception of time implies the co-existence of multiple existents. These existents are connected by the laws of nature, which constitute their simultaneous existence.

Recall in this context that I have rejected the idea of simultaneous laws of nature. If two existents, *x* and *y*, are connected by the laws of nature, this connection can only be manifested at a later moment in time. Hence, this description relies on the assumption that some connection can be found between *x* and *y* and a later existent *z*, which can be said to be determined by these previous existents. Hence, this idea ipso facto relies on the possibility of identifying a connection between existents at different times, which constitutes a continuation or a (weak) form of identity in time. That is, it allows to us to identify that *z* is, for example, determined by *x* and *y*. In this case, *z* can, according to the circumstances, be either identified with *x*, *y*, their merger, or as a distinct entity that is determined by them.

Although this conclusion suffices for demonstrating that some form of a connection of continuation between existents at different times must be identifiable, it is important to recognize that the conceptual apparatus that I have developed here requires an even stronger connection. The conceptual apparatus that was uncovered so far does indeed imply that there must be some form of strong identity in time, according to which an existent should be identified as persisting in time. Furthermore, coupled with the idea that the laws of nature connect multiple occurrences at one time with an occurrence in other times, this implies the idea of influence, that is, not only *y*, which exists simultaneously with *x* at time t', can be said to continue to exist at a later time t'', but it can also be said to be determined by *x*. Furthermore, once this idea is introduced regarding one entity, it can apply to multiple entities, which implies the idea of reciprocal determination between entities that persist in time.

To see how this idea is implied by the conceptual apparatus that has been developed so far, recall that I have argued that to allow for the ascription of different truth-values to the same propositions without a contradiction, it is necessary to make the distinction between a proposition in language and what this proposition describes, that is, between ‘Nφ’ and Nφ. Furthermore, this distinction equally applies to propositions that describe the past, for example, Pφ. The distinction between a description of the past and what it describes is demonstrated by the possibility of a non-inferential description of the past. Non-inferential descriptions of the past imply the possibility that in the past the proposition Nφ was true and was accepted as such in the conceptual scheme, that is, P(N(N$\varphi\wedge$‘T’ ‘Nφ’)). The realization of the property φ in the past is connected by the laws of nature to its description in the conceptual scheme in the present.

Hence, there must be at least one case in which a continuous entity, that is, the conceptual scheme, has been determined by a simultaneous entity. Furthermore, the conceptual scheme is not subject to any unique rules. It exists in the same sense as any other existent in nature and is subject to the same laws of nature. Hence, this possibility applies to anything in nature. There must therefore be a way of determining identity through time.

It is yet to be explained what allows the identification of such a connection between existents at different times. In our conceptual scheme, this is provided by the idea of space, which has yet to be introduced into the conceptual structure I am developing here. This is mediated by the idea of a location in space, which enables determination of the existence of a connection between existents in time. Thus, the identity in spatial location, or proximity in the case of moving entities, established a connection between existents at different times.

My analysis hence implies the existence of a relation of determination between different times, which allows establishing a connection between existents at different times. However, it also implies the existence of a relation between simultaneous existents. Recall that the laws of nature tie together simultaneous existents, thus constituting the relations of simultaneity, by integrating them into a relation of reciprocal determination. Because the laws of nature cannot be global, this connection between simultaneous existents necessarily singles out some of the existents at a certain time, thus enabling the introduction of an idea of locality. The idea of locality is used in this context to represent the relation between existents which reciprocally determine one another. This idea will be developed in the next section.

However, these local connections cannot be isolated from one another. The unity of nature requires that these laws, which tie only some of the entities that exist simultaneously in connections of reciprocal determination, ultimately unify all the entities in nature. That is, the laws of nature must connect entities that do not directly determine one another. Hence, these connections must constitute some order between simultaneous existents, which reflect their reciprocal determination according to the laws of nature.

Thus, the conclusion of my analysis leads to a primitive, yet unstructured, idea of space. This idea reflects the reciprocal connections that obtain between all the entities that exist simultaneously. Although this connection is constituted by the laws of nature, which connect occurrences at different times (and thus constitute time), these laws not only constitute the temporal order between occurrences, but also constitute the relations of co-existence in time, that is, space.

To reiterate, I have shown that some form of connection between existents at different times must be identifiable; that a general distinction between local and global should be introduced; and that there must be some form of order between simultaneous existents. All these are implied by the laws of nature, which connect both simultaneous and successive occurrences, thus constituting the unity of nature. Thus, all these ideas, which reflect different relations of determination, are found to be mutually linked, and imply some sort of order between existents. These forms of an order will be developed and explored in what follows into a conception of space.

6.6 Locality

The analysis of the last section introduced another property into the conceptual structure that constitutes nature, that is, relations between existents, which reflects the order of their reciprocal determination. This is the property that will be developed in what follows into the idea of apace. Furthermore, these relations will extend the narrow circularity discussed at the beginning of Sect. 6.5, between time and the properties that change in time. The changing properties in time will hence be identified as spatial properties.

In this section, I wish to explore the notion of locality. To begin with, it is important to understand the connection between the laws of nature and the idea of locality. I wish to emphasize that I do not argue that, by connecting simultaneous existents, the laws of nature give structure to an otherwise unstructured pre-existing manifold. Firstly, this is because the relation of simultaneous existence itself is given meaning by the laws of nature. Moreover, the laws of nature are not a separate organizing principle, which forces order on separately existing multiple entities. As I have argued, the laws of nature reflect the conceptual structure of time, which itself is the most fundamental property of all existents, which constitutes all existents in nature.

These conclusions give rise to an important insight. I argued that the change in time, which is described by the laws of nature, is constituted by the reciprocal determination of simultaneous existents. This gives rise to spatial properties. If these are the properties that change in time, then the change in time is constituted by a change in space. A change in space, coupled with the idea of identity in time, gives rise to the idea of movement in space.

However, the conceptual structure I am developing here still lacks any structure that supports the idea of movement in space. Much like the idea of a change in time, this idea relies on the ideas of positions and distances in space. Without the ability to apply these distinctions, it is impossible to employ the notion of change in space in descriptions of nature.

To develop these ideas, I first wish to further explore the need for a structured co-existence. To begin with, it is important to revisit the rejection of global laws of nature, and understand the significance of the idea of local laws of nature. Recall that I have argued that the laws of nature cannot be global, in the sense that it is impossible to consider all that co-exists at a certain time while determining what exists at other times. This is not merely an epistemological limitation of human agents. Rather, this is a necessary limitation that arises from the fact that nature is described from a limited point of view that is internal to it, which is constitutive of nature. The significance of this fact will be explored and explained thoroughly in Chap. 7. Meanwhile, it suffices to note that this does not represent a contingent limitation, but is rather a necessary feature that is a presumption of any objective description of nature.

In Sect. 5.14, I argued against the idea of global laws of nature. However, my discussion was limited because at that time the concept of space was yet to be introduced. Hence, my discussion focused on the restrictive conditions on any possible identification, and hence use, of the laws of nature, for formulating an objective description of nature, from the limited point of view of a conceptual scheme. Although these considerations are valid, they do not represent the deepest insight into this subject. To understand the true significance of the locality of the laws of nature, it is necessary to delve into the essence of space, and how the laws of nature shape its structure.

I wish to turn the attention again to how non-inferential descriptions of nature are made possible. Recall that these descriptions are formulated by reliance on the laws of nature, which connect occurrences at different times. Although the laws of nature are symmetrical, I have argued (in Sect. 5.7) that knowledge of the future cannot be analogous to the knowledge of the past. This asymmetry will be explained in Chap. 7. In the present context, what is relevant is how earlier occurrences determine later occurrences.

Because nature is described from the point of view of a conceptual scheme, whose default context operator is the present, the past must be available for a full reconstruction from the present. That is, the laws of nature must be such that they make it conceptually possible to formulate a complete description of the past. As already mentioned, the conceptual scheme necessarily offers a limited view of nature. Hence, a single conceptual scheme cannot formulate a complete description of the history of all nature. However, it should be possible for the conceptual scheme to reconstruct at least some of the conditions that determine its current state.

In light of these reminders, suppose that nature is constituted by a global law of nature. To begin with, a law that connects all the occurrences at one time with all the occurrences at another time is unable to explain how a conceptual scheme can formulate even a limited description of the past. Such a law makes no distinctions between the different determining conditions at each time, and hence cannot enable reconstruction, based on a limited description of the present, which is available to the conceptual scheme, of any form of description of the past, whether complete or partial.

To allow creation of a description of the past, even if a limited one, based on a limited description of the present that is available to the conceptual scheme, a global law of nature would have to consider all past occurrences in determining any occurrence at a certain time t'. A conceptual scheme is itself part of nature, and hence determined in the same way as other parts of nature. That is, the state of a conceptual scheme is determined by the totality of everything in nature prior to time t'.

To begin with, it might be thought that if all the things in nature take part in the determination of any future part of nature, then it follows that all other parts of nature are determined in the same way. If any part of nature is determined by the same law, according to the same previous conditions, then everything that is so determined must be identical. However, this absurd conclusion can be avoided by rejecting the presumption that the conditions that determine each part of nature are identical. Admittedly, I have argued that simultaneous existence is established according to participation in the mutual determination of future occurrences. However, does not imply that there is a global relation of simultaneity. The occurrences that are determined as simultaneous from the perspective of one point in space may be different from the occurrences that are determined as simultaneous from the perspective of a different point in space. Hence, a global law of nature does not necessarily imply that all future occurrences are determined by the same factors.

A more problematic implication of the suggestion that all preceding occurrences take part in the determination of the conceptual scheme is the fact that it makes it impossible to describe the past. Recall that an objective description of the past must be given from a point of view that is located in the present. This is the only way to objectively describe the past, and hence a full description of the past must be available from the present. This does not imply a single conceptual scheme which includes all that is needed to describe all past occurrences in nature. It is nevertheless necessary that it would be possible to infer, based on the current state of the conceptual scheme, or any other part of nature, at least some, if not all, the previous simultaneous conditions that determined this state. This is the only way that allows formulating a description of the past from an internal and limited view of nature. However, this is impossible if all the preceding occurrences in nature take part in determining the current state of each part of nature. This would make the current state of the conceptual scheme a necessary and sufficient condition for everything else that existed before it. Given the structure of the laws of nature, this would also make the current state of the conceptual scheme a necessary condition for all past occurrences, and hence would imply that the conceptual scheme is all that exists in the present.

One might try to evade this conclusion by making distinctions in the way in which past occurrences determine the current state of the conceptual scheme. I have so far discussed the case in which the contribution of all the preceding occurrences, for determining the current state of the conceptual scheme, is identical. However, it might be claimed that, even if all the preceding occurrences take part in determining the current state of the conceptual scheme, their contribution may not be identical. For example, some occurrences may have a lesser contribution to the determination of the current state of the conceptual scheme, or indeed to the determination of any future occurrence φ. Furthermore, the contribution of some components could be so small as to be ignored. This can explain how such global laws of nature can be possible, even though they incorporate the totality of occurrences at each time.

In answer to this suggestion, it should first be noted that this distinction, regarding the weight of each occurrence in determining a future occurrence, can be seen as constituting an order in the manifold of simultaneous occurrences at each time. Thus, this suggestion not only constitutes space, but also allows giving meaning to a notion of "locality." According to this suggestion, occurrences that have more influence on determining a future occurrence φ are identified as located closer to it in space. Thus, the structure of space is constituted by, and thus reflects, the lawful connections between occurrences.

Although this suggestion can allegedly explain the existence of space, and its structure, I do not believe that this is a correct description of the structure of the laws of nature. The reason for its failure is similar to that of global laws of nature in general; that is, the fact that all occurrences at a certain time are alleged to take part in the determination of any single occurrence at a later time. This suggestion may solve the *pragmatic* problem of relying on the laws of nature for formulating descriptions of the past. That is, it may allow the formulation of relatively accurate descriptions of the past. However, the problem remains, that it implies that the current state of the conceptual scheme is a sufficient condition for all past occurrences.

The nature of locality should by now be clear. Locality reflects the limited availability of nature as a whole from the point of view of the conceptual scheme. This is not unique to the conceptual scheme, but rather characterizes the relation between any part of nature and nature as a whole. This constitutes the limited point of view of the conceptual scheme over nature. Had all parts of nature been taking part in the determination of future occurrences, it would have been impossible to make the distinction between the manifold of occurrences at a certain time. This would eliminate all the distinctions between the occurrences that exist simultaneously, as with the conceptual scheme and what it describes.

This conclusion coheres with the fact that the conceptual scheme must reflect an internal and limited point of view of nature. Given that this is an internal and limited point of view, it necessarily reflects only part of nature. Furthermore, given that the laws of nature connect the occurrences in nature, thus constituting the unity of nature, it should be clear why the laws of nature cannot be global, and the immediate connections between occurrences must be limited in scope. While the unity of nature requires all the occurrences in nature to be connected at some point by the laws of nature, the distinctions between the various existents require their separation by the delayed temporal determination. The significance of this fact will be better understood in the next chapter.

In conclusion, locality reflects the fact that simultaneous existents are separated by their reciprocal determination, and hence action-at-a-distance is impossible. Proximity in space echoes a more immediate determination of future occurrences. Connections of reciprocal determination can, therefore, only be local.

6.7 The Constitution of Space

The previous section concludes that the idea of locality is fundamental for constructing an objective conception of nature from a limited point of view that is internal to nature. However, the idea of locality has no meaning unless the manifold of simultaneous existents is structured. As yet, it is unclear how the idea of space can be developed (rather than artificially introduced) based on the conceptual structure I have developed so far.

The starting point of my current inquiry is the reminder that was given in Sect. 6.5, according to which time requires a continuous point of view from which nature is described. The continuity of this point of view allows it to record the changes in time. This is done with the help of a conceptual scheme, which keeps its identity in time. In the context of a conceptual scheme, this identity implies the preservation of the conceptual structure, truth-values of propositions, and documentation of the history of truth-value ascriptions, that is, a description of its own changes in time. All these were found necessary to explain the ability to describe nature with the help of contingent propositions.

My analysis of the possibility of describing the past has shown that the laws of nature are necessary for preserving the conceptual scheme through time. However, being connected by laws is not a sufficient condition for identity through time, in our complex conceptual scheme. Normally, determining identity in time requires not only a continuous existence in time, but also a continuous existence in space. That is, a necessary condition for the identification of an entity φ at time T and spatial location S, with entity an ψ at time T' and spatial location S', is the continuous existence of φ between time T and T', and between S and S′. However, it might be argued that this condition for identity only arises when there is a possibility of multiple entities of a certain type. For example, if multiple atoms exist at times T and T', it is necessary to rely on space, which allows distinguishing between the different atoms at each time, and tracing a continuous existence in space can determine the identity of atoms at different times. Similarly, the same question can arise in the case of a multitude of conceptual schemes, and in this case, the spatial criterion is crucial. It might be argued, however, that if the existence of multiple entities of a certain type is impossible, the spatial criterion has no meaning. Furthermore, the possibility of multiple conceptual schemes has yet to be discussed, and hence the idea of spatial location was not shown to be necessary for identity through time.

I have argued that the objectivity of time requires giving meaning to the idea of a temporal location. The challenge is now to show that objectivity also relies on the idea of a spatial location.

Before giving a justification for this claim, I wish to warn against a false consideration that might lead this inquiry astray. The previous analysis has shown that time implies the existence of laws of nature that tie simultaneous occurrences by reciprocal determination in time. It might be argued that a reciprocal determination in time can only take place if the involved parties persist in time. For an entity φ to determine a simultaneous entity ψ, given that the laws of nature connect occurrences at different times, ψ must persist in the future. This suggestion can be represented, in Kantian terms, as an attempt to infer the "First Analogy of Experience" based on the conclusion of the "Third Analogy of Experience" (Kant, 1929, pp. B224–B233, B256–B262).

The flaw in this suggestion is that it interprets the findings of Sect. 6.5 in strong terms of identity through time. Although there must be some form of connection between the occurrences at different times, it is not clear that this connection must be that of strong identity, which implies the existence of substances (as persisting entities). My analysis indicated the existence of one entity that requires a continuous existence in time, that is, the conceptual scheme, but no more.

The reason why it is necessary to determine locations in space, and reidentify locations in space at different times, is the need to identify objective changes. The first consideration that is relevant in this respect is similar to the one that underlies Kant's argument in the "Second Analogy of Experience" (Kant, 1929, pp. B232–B256). To understand the need for the identification of locations in space, it is important to remember that, due to the objectivity of time, the identification of changes is not straightforward, and relies on a complex conceptual structure. To see why, I rely on a minimalist definition of a change as a difference in the truth-value of a proposition of the type "φ exists" at different times. Prima facie, this definition allows a straightforward identification of change. However, this is only true regarding the subjective realm, that is, a change in the conceptual scheme itself. Regarding the objective realm, in which the possibilities of mistake and ignorance are relevant, this is not the case. In the present context, it suffices to rely on the possibility of ignorance, to show that a change in the truth-value that is ascribed to the proposition "φ exists" in a conceptual scheme does not imply an objective change. Thus, if φ is recognized at time T but not in time T', it is always possible that it nevertheless exists without being noticed. Hence, the ascription of different truth-values to a proposition does not imply an objective change in nature.

Interestingly, to make this claim, Kant relied on the possibility of changing the point of view on the world. Kant's famous example is that of first perceiving the rooftop of a house and then its lower part (Kant, 1929, p. B237). This is an example of a subjective change, that does not imply an objective change in what is described, that is, the state of the house. However, this idea of a change in point of view presupposes spatial distinctions, and hence far exceeds the conceptual structure that was developed in the previous analyses. Furthermore, because I am trying to show that space is necessary for making objective time determinations, any reliance on this line of reasoning would beg the question.

To determine whether two different descriptions of nature record an objective change, the ability to identify locations in space becomes crucial. Due to the limited point of view of nature, there must be a way of determining that φ is no longer realized, rather than not recognized. If there is no way to make distinctions in the manifold of existents at time T', it is impossible to determine that φ is no longer realized. Only if it is possible to determine that these descriptions apply to the same place at different times is it possible to infer that a change had indeed occurred.

A possible objection that might be raised to my claim, that the reidentification of locations in space is necessary for the identification of changes, is that I have not developed the idea of a spatial location based on the conceptual structure I have introduced so far. I presented a difficulty, that of identifying objective changes, and showed that the idea of positions in space can solve this problem. I have not shown how this solution arises from the conceptual apparatus that has been developed so far, and hence, even if the idea of spatial position can solve this problem, it does not show that it is a necessary feature of nature. There might be, so this objection continues, another possible solution, or solutions, for this difficulty.

This objection is, I believe, justified. To answer the challenge, I shall rely on the reminder that objectivity is constituted by the laws of nature. If locations in space are needed, they must be needed to identify the laws of nature.

The laws of nature connect occurrences at different times, such that they are sufficient and necessary conditions for one another. However, there is a multitude of occurrences at each time, and although it is trivial to claim that all the occurrences at one time T determine all the occurrences at future time T', I have argued against the idea of global laws of nature (in Sects. 5.14 and 6.8). Such a global law, which relies on everything that exists at one time determining all that exist at another time, is unable to fulfill the role it was introduced for, that is, to allow us to create an objective account of nature.

An important implication of the impossibility of global laws of nature is the need for an idea of locality, that is, to make distinctions between all the conditions at a certain time T, and determine which are relevant for the determination of a future occurrence at time T', and identify which occurrences they determine. Furthermore, the identification of the relevant conditions must be available from the point of view of the conceptual scheme from which nature is described, and this point of view is necessarily partial.

Furthermore, while determining the relevant conditions, it is important to remember the unity of nature, that is, the requirement of all existents in nature to be connected by the laws of nature at one time or another (discussed in Sect. 5.11). Hence, it is not sufficient to distinguish between the different conditions at time T and determine which of them are relevant to the determination of a specific occurrence at a later time T'. It is also necessary to establish which conditions are relevant to determine a specific occurrence at an even later time T''.

I have described this difficulty in epistemological terms, as a problem of identification. However, this is not an epistemological problem, but rather a conceptual problem. To solve this problem, it is necessary to explain how the objective description of nature is possible. The answer, therefore, lies in the laws of nature themselves, which constitute objective temporal distinctions. I have already shown that they determine distances in time between occurrences, that is, whether occurrences follow immediately one after the other or are separated by a temporal interval. However, they create the connection between occurrences at different times, and this connection constitutes an identity in spatial location, that is, the uniqueness of some occurrences in the manifolds of simultaneous existents. Locations in space are therefore determined, first and foremost, in relation to locations at different times.

That is, the basic notion is that of identity in locations at different times, which is determined by the laws of nature. There is no meaning to the idea of a spatial location at a certain time unless it is an abstraction from the relation of identity to locations at other times, whether preceding or succeeding.

The manifold of existents at different times is therefore connected at specific points to one another. These points of connection are established by the laws of nature, and constitute an identity in temporal location. Although, while considering the connection between the manifolds at two close points in time, the laws of nature only establish limited connections between isolated existents at each time, the laws of nature are supposed to ultimately unify all occurrences in nature. Hence, the larger the distances in time between the manifolds of occurrences, the more connections can be found between different times. This is the basis for the constitution of space.

Before concluding this section, I wish to return to the epistemological question. As previously mentioned, the conceptual and the epistemological questions are, in this context, the same. The ability of the conceptual scheme to determine identity in time and that there is a connection between occurrences at different times—two issues that are two sides of the same problem—is found in the explanation for the same conditions that give rise to this problem. The fact that the conceptual scheme describes what exists from a limited and internal point of view to nature led me to introduce the laws of nature. These laws of nature constitute the objective realm, and connect in time the occurrences in nature. These laws also connect occurrences to the conceptual scheme, and this explains, not only how a conceptual scheme can describe nature, but also gives structure to the connections between the conceptual scheme and what it describes. Since the conceptual scheme itself is part of this structure of laws, what is available from the point of view of the conceptual scheme is not random and unstructured, but is rather reflective of the location of the conceptual scheme and its connections with adjacent locations in space.

6.8 The Structure of Space

The analysis undertaken in the previous section has shown the importance of a structure for the multitude of simultaneous occurrences. The basic concept in this structure is the idea of a location, or a position, among all the existents at a certain time, that is, in space. Hence, this idea requires giving a structure to the manifold of simultaneous existents. However, the idea of a location is not only relative to the totality of simultaneous existents. It is first and foremost an indication of a connection of determination between existents at different times. That is, it has no meaning as a relation to simultaneous existents unless this reflects connections to positions at other times. These connections of determination are established by the laws of nature. Thus an occurrence, or occurrences, that determine another occurrence at a different time are identified as located at the same, or adjacent, position in space.

Locations in space, therefore, denote points in which correspondence is established between existents at different times.

Positions and spatial relations are interdependent. Location is not determined in relation to absolute space, but rather in relation to other simultaneous occurrences, that is, by spatial relations. However, this does not contradict the previous claim that position has no meaning except in relation to positions at other times. This is because spatial relations themselves are determined by their participation in the determination of occurrences at other times.

At this point, it might be helpful to compare a position in time with a position in space. Acknowledging the analogy between the two notions of a location in a manifold of existents can help us to better understand the connection between time and space. To begin with, locating a position in time allows determining identity in temporal location, that is, a relation of simultaneity, between two existents in different spatial locations. Locating a position in time also allows determining the distance to other positions in time, and thus identifying the lawful relations that obtain between them. Analogically, locating a position in space allows determining identity between two existents at different times, thus establishing the lawful relation between them. I should stress, however, that this analogy does not imply a symmetry between time and space. Time and space fulfill different roles in the laws of nature, which constitute both time and space. Although the laws of nature connect existents in adjacent locations in time and space, at the fundamental level the laws of nature describe a connection between occurrences at different times, thus constituting time, while space is a corollary of this basic relation.

The idea of a location in time implies spatial order. This is because I have rejected the idea of isolated chains of connections between occurrences, and argued that nature must be unified by the laws of nature (in Sect. 5.11). The connection between locations and the structure of space can be uncovered by exploring the implications of the idea that a location in space has a temporal meaning. The identification of a location in space is, first and foremost, a determination of the identity of spatial locations at different times. This identification is based on the laws of nature, which connect locations at different times by way of reciprocal determination. Because all the occurrences are part of unified nature, they must be connected by the laws of nature. These laws connect different times, and hence all occurrences must be connected by the way they determine occurrences at other times. This implies that there must also be a connection between simultaneous positions in space.

However, I have argued that global laws of nature are impossible, and hence laws of nature must be local, that is, connect only some occurrences, which are thus deemed to be closely located in space. This implies that it is impossible for all occurrences at a certain time to determine together any future occurrence. Instead, there must be a principle that determines an order in which occurrences take part in determining future occurrences. The closer occurrences are in space, the closer in time the occurrences that they determine. That is, only occurrences that are located close to a location L at time T determine what happens immediately afterward, at time T', at this location. The spatial order of occurrences around location L reflects the relevance of these occurrences to the determination of later occurrences at this location. That is, the later the occurrence that is determined at a later time at location L (that is, at time T'' > T'), the more occurrences take part in its determination, and hence the further they are in space from location L.

The structure of space thus reflects the relations of determination between occurrences at a certain point in time. Notice that this gives precise meaning to the idea of locality, which until now was only presented in contrast to the idea of global laws of nature. It again reflects the connection between the laws of nature and locality, which explains why the laws of nature must be local—it is the laws of nature themselves that give meaning to relations in space, and hence give meaning to the idea of proximity in space. The locality is defined precisely by the ability to determine occurrences at other times according to the laws of nature.

Further, notice that the later the occurrence that is determined, the more occurrences are relevant for its determination. This is necessary to guarantee the unity of nature, that more occurrences take part in determining other occurrences. However, I have also argued that the impossibility of global laws of nature resides in the fact that it does not allow distinguishing between the different factors, or existents, that determine a certain occurrence. If too many occurrences take part in the determination of future occurrences, it is impossible to identify them individually. This does not allow the formulation of an objective description of the past, and hence undermines the conceptual structure that was found to constitute time.

This condition may lead some to the conclusion that an entity, for example, the conceptual scheme itself, can only be determined by a single previous condition. For if multiple conditions would together determine the state of the conceptual scheme, it would be impossible, so it might be argued, to subsequently identify them individually. However, this claim ignores some of the conclusions of my analysis. First and foremost, it ignores the idea of size. For I have argued that entities cannot occupy the same location in space. Furthermore, the relations of the entities in space reflect the order of their engagement in determining future occurrences. Thus, when considering the conditions that determine a later occurrence, the later this occurrence, the wider the area of the determining conditions. Hence, the bigger the size of the determined entity, the more previous factors take part in its determination. Thus, the bigger the conceptual scheme, the more it is determined by previous conditions, but the more it is able to identify these conditions individually. Hence, a conceptual scheme cannot be a point-like entity.

It is now possible to show that a spatial location is a distinguishing feature, and hence more than one entity cannot be simultaneously located in a certain position in space. The reason why two separate entities cannot occupy the same location at once is that they are supposed to be distinguishable. Otherwise, it would be impossible to determine that there are two of them. However, since spatial location reflects a relation of determination of other entities, if two entities had been located at the same position in space, it would have been impossible to make any distinction between them in any objective description of nature. This consideration also explains the significance of positions in determining relations of identity, or non-identity, through time.

The final question that is left to explore is the actual necessary structure of space. Is the structure of space necessarily three-dimensional, and if so, why? What is the geometry of such space? These and similar questions cannot be addressed in the confines of this inquiry. My current task is more modest. I wish to pave the way for future exploration of this complex subject, by illuminating some of the considerations that ought to govern such an inquiry.

The starting point of any inquiry into the exact structure of space must begin with the structure of time, which, as I have developed it in Chap. 4, is one-dimensional, in which every two occurrences are either simultaneous, or where one precedes the other. Notice that this fact alone determines the general structure of space. For recall that my analysis showed that time is more basic than space, and hence constitutes space. The temporal order to occurrences in space has a temporal meaning—it reflects the temporal order in which they engage in the determination of occurrences at other times. Hence, space is constituted by a structure that is analogical to the structure of time.

The importance of this insight cannot be overestimated. This analogy between time and space was noticed for generations and puzzled philosophers and physicists. It was correctly thought that this cannot be a mere coincidence, and hence an explanation for this fact was sought. Finally, the analogy between time and space has allowed modern physicists to interpret nature as 4-dimensional, and for some to reject the duality between time and space. According to this position, there is no need to explain the similarity between time and space—they are the same. Rather, what should be explained is why time seems to be fundamentally different, while it is essentially not different from the three dimensions of space.

My analysis, on the other hand, shows that time is more basic than space. The structural analogy between time and space, however, offers an explanation that relates both to the connection between time and space, and to the principles that establish the structure of space. As I have explained, time is the most fundamental property of existence, and hence constitutes space. The temporal order of occurrences in space has a temporal significance—it is the temporal order in which they engage in the determination of occurrences at other times. The structure of the dimensions of space is analogical to the structure of time because they are constituted by temporal distinctions.

Although this consideration offers an important insight into the structure of space, it is only the beginning of the inquiry. For space is seemingly three-dimensional, and if it is accepted that this (empirical fact) indeed reflects its true structure, the question of why this is so and whether it is a necessary feature of nature remains. Furthermore, due to the close relationship between time and space, it seems strange that while time is (necessarily) one-dimensional, space is three-dimensional. However, exploring these and other questions about the structure of space extends beyond the scope of this inquiry. Nevertheless, I trust that the starting point for this inquiry, as well as the guiding principles for its undertaking, have been established here.

6.9 Speed of Determination and the Measurement of Space

In the last section of this chapter, I wish to lay down the last element in the foundations of space and time. The end of the philosophy of nature is the exploration of the fabric of time and space, as the basic constituents of existence. In the current inquiry, I concentrate on the interrelation between these fundamental features of nature, which is reflected, as shall be soon apparent, in the idea of a change in space. Understanding this connection between time and space will also allow me to introduce the final element in the constitution of space, that is, the tools for determining distances in space, in analogy to those that were introduced for determining distances in time. These tools are necessary for any investigation of nature, whether empirical or conceptual, for they explain the ability to apply mathematical relations in descriptions of time and space, thus enabling the formulation of quantitative theories of nature.

Recall that spatial relations must encompass all existents in nature in a single unified structure. This is an implication of the demand for a unity of nature, which is realized by the laws of nature, and reflected in the structure of space. Because the laws that establish these relations are temporal laws, this structure is of temporal meaning. Indeed, space has been found to give meaning to the idea of a change, thus allowing escape from the narrow circularity that is involved in characterizing time in terms of change. Furthermore, what constitutes space are the relations between existents. Hence, it follows that spatial relations must themselves be subject to temporal determinations, and hence can either stay constant or change in time.

To reiterate, time is constituted by laws of nature, which describe connections between existents at different times. These laws of nature describe the change in time. They not only determine the order of existents in time, but also order in space, that is, relations between simultaneous existents, thus constituting space. Hence, the temporal laws of nature describe the change in the spatial relations, that is, a change in positions in space. A change in position is movement in space.

The laws of nature, which constitute time, also provide the means for determining distances in time—a necessary requirement for establishing objective temporal relations. As was explained in Sect. 6.4, time is measured by the use of regularities, which offer a standard unit for measuring time. Although the choice of using any particular regularity in nature as a unit of measurement may be found to be practically adequate, there must be a fundamental change in nature that is uniform and accurate by definition, as a necessary feature of nature.

My analysis of spatial relations has shown that the change that constitutes time is constituted by the change in spatial relations. Spatial relations reflect relations of determination in time, and hence are relative to positions in time. Take for example an entity E that is located at position P at time T. The spatial relations of other existents, relative to P, reflect the order of their determination of E. The state of entity E at a later time, T', is determined by the neighboring entities, according to their spatial position. The farther a particular entity is from the location P, the longer it takes for it to take part in the determination of E. Thus, in time and as the neighboring entities determine each other, their spatial relations must change to reflect this change. The change in these relations is a change in the spatial relations, that is, movement.

The change that gives meaning to the measurement of time is determined by the laws of nature, and is constituted by the change in spatial relations, that is, movements in space. Since spatial relations are determined according to the participation of the entities involved in the determination of the changes in time, which provides the standard for determining distances in time, spatial relations are also measurable according to their relations to changes in time. In other words, the change in positions in space, that is movement, is measurable by the standard for determining distances in time.

This conclusion should not come as a surprise. For I have argued earlier that spatial relations are constituted by temporal relations, which are the most fundamental features of existence. If time is measurable, so is space. Furthermore, the change in time is constituted by the spatial relations, and hence these changes are also measurable by the temporal standards of measurement.

A consequence of these considerations is that the unit of measurement of time is determined by movement in space. That is, movement in space must be at a constant temporal rate. Admittedly, the rate of temporal change is measured by the change in space, that is, distance in space. However, distances in space, as I have argued, are themselves constituted by temporal order, and hence measured by time.

There is an element of circularity in my description of these relations which is not, however, a vicious circularity. It merely implies that there is a close relationship between time and space, which is expressed in the idea of a change, which unifies the two notions together. In simple words, it states that there is a fundamental and constant change in nature, which provides a unit of measurement for both space and time. Given any arbitrary unit Δt for the measurement of time, it corresponds to a unit of space Δs, which is the distance in space covered by the speed of determination at Δt. Thus, this conclusion not only gives meaning to the idea of objective time measurement through a change in spatial relations, but it also gives meaning to the idea of objective measurement of space.

The speed of determination, which is always constant, must be finite. This is implied not only by the need to provide a unit for the measurement of both time and space, which cannot be fulfilled by an immediate determination. It is also implied by the notion of spatial relations, which are constituted by order of determination, and follows directly from the conclusion that the relation of determination must be local.

The implications of the claim that the speed of determination is fundamental are twofold. To begin with, it implies that no change in nature can surpass the speed of this fundamental determination. This is because spatial relations are constituted by relations of determination. Any alleged speed that exceeds the fundamental speed of determination undermines the spatial relations which give meaning to the measurement of space, and hence also to the measurement of speed.

The other implication is that there cannot be a *fundamental change* in nature that is slower than this speed. Again, the explanation relates to the fact that spatial relations reflect the temporal order of determination. Just as it is impossible to exceed the fundamental speed in nature, the fundamental change in nature cannot be slower. However, this does not imply that every change in space must be of this fundamental speed. Although at the fundamental level all changes necessarily take place according to the constitutive speed of determination, non-fundamental entities can move at a slower speed (although not at a greater speed).

It might be thought that a distinction should be made between entities and the speed at which these entities determine one another. I should stress, however, that the temporal-spatial structure of relations that I have described above is a necessary feature of existence, and hence describe the structure of any entity in existence. Some entities do move slower than the speed of determination, these, however, are necessarily high-order entities. Notice, further, that my analysis has implied the existence of at least one such entity, that is, the conceptual scheme, whose complex structure is realized as part of nature.

Before concluding this chapter, I wish to point out some important implications and insights that follow from my analysis. The most important implication of this analysis regards the essence of the entities that exist in space and time. As should be obvious by now, what constitutes existence is changes in time, which consists of spatial relations entities, which themselves reflect relations of determination, that is, of change, in time. These temporo-spatial relations are a necessary feature of everything that exists—they constitute existence per se. To say that something exists is to say that it exists in time and space as a determining factor. What I have called occurrences or, more generally, entities, are nothing but determining factors. Their existence consists of nothing more than their reciprocal determination in time.

Notice that this conclusion does not rule out the possibility of empty space. The fact that the location of entities reflects not only an order among all other simultaneous entities, but also the time that it takes for one occurrence to determine another, can explain the possibility of empty space—a space in which there are no determining factors.

Furthermore, it is now possible to better understand why two entities cannot occupy the same spatial location. Previously, I have argued that they would be indistinguishable. However, now it is possible to achieve a deeper insight into this fact. The reason why two entities, which are determining factors, cannot occupy the same spatial location is that they would have to determine one another instantaneously. However, "two entities" that determine one another instantaneously are ipso facto identical to one another.

My intention in this chapter is merely to expose the basic structure of time and space at the most fundamental level. I believe that this task has been completed. I shall therefore leave any further investigations of this subject for future research. The task I now turn to is offering a better understanding of the meaning of this fundamental conceptual structure of nature.

References

Dennett, D. (1991). *Consciousness explained.* Allen Lane.
Goodman, N. (1983). *Fact, fiction, & forecast* (4th ed.). Harvard University Press.
Hume, D. (1978). In L. A. Selby-Bigge (Ed.), *A treatise of human nature*. Oxford University Press.
James, W. (1890). *The principles of psychology*. Henry Holt.
Kant, I. (1929). *Critique of pure reason* (N. Kemp-Smith, Trans.). Macmillan.
Le Poidevin, R. (2004). A puzzle concerning time perception. *Synthese, 142*(1), 109–142.
Mellor, H. D. (1998). *Real time II*. Routledge.
Strawson, P. F. (1959). *Individuals: An essay in descriptive metaphysics*. Routledge.

Chapter 7
The Constitution of Nature

Abstract This chapter is dedicated to better understanding the conception of nature that was introduced in the previous chapters, which I term "internal relationalism." I begin by explaining the implications of the conclusion that existence is a relation. I then explain the meaning of the laws of nature and how they constitute time and space. Following this, I return to the notion of time and rely on these insights to explain the arrow of time. Finally, I show how the ancient riddle "why is there something rather than nothing?" is answered by the internal relationalism conception of nature, by showing that it implies that nature necessarily exists.

Keywords Arrow of time · Existence · Internal relationalism · Laws of nature · Structure of nature

7.1 Introduction

In this penultimate chapter of this book, I wish to explore the implications and insights of the conception of nature that was developed in the previous chapters, that is, internal relationalism. I do not intend to further enrich the complex conceptual structure that I have introduced in the previous chapters, but merely to offer a better understanding of its significance. In what follows, I shall focus on the most important aspects of this structure, which relate to some of the most fundamental questions that have plagued humanity for millennia: What is existence? Why do things change in nature? What is the origin of nature? What explains the flow of time? And finally, the most important question, whose answer encapsulates the deepest understanding of nature, that is, why is there something rather than nothing?

G. Yehezkel, *Conceptual Principles of Natural Philosophy*, Synthese Library 495, https://doi.org/10.1007/978-3-031-82108-0_7

7.2 Existence as a Relation

The starting point of my metaphysical inquiry was an analysis of the concept of existence. My analysis has shown that the conceptual basis of the concept of existence is the category of "contingent entities," that is, entities whose existence is described by contingent propositions. The inquiry that followed was dedicated solely to uncovering the conceptual structure that supports the contingency of propositions in language, thereby exposing the true meaning of the concept of existence.

My analysis has shown that contingent propositions can be used for describing nature only if this description requires the same propositions to be both true and false, and that this is possible without contradiction only if nature is described by the use of temporal indexical expressions. These indexical expressions were found to express the changing and internal point of view of nature. This relation, between the point of view and what it describes, was found to involve not only temporal relations but also spatial relations. These temporo-spatial relations are the fundamental conceptual structure that supports the concept of existence.

The point of view of nature is described by a conceptual scheme. This conceptual scheme stands in similar temporo-spatial relations to the things it describes. It is these temporo-spatial relations, between the conceptual scheme and nature, as between any other existents, that constitute existence.

A conceptual scheme can be seen as an abstract structure of meaningful relations. However, in this case, it was found to be necessarily embodied as an entity in nature. The temporo-spatial relations are not only the relations that obtain between this entity and other existents. They are the relations that obtain between any existents in nature. Furthermore, these relations are not only relations that obtain between existents. They are necessary features of existence, and hence constitute any existent in nature.

This is the meaning of the conclusion that existence is a relation. Existence is constituted by pure temporal-spatial relations. Any property that exists in nature is a relation, and constituted, at the most fundamental level, by temporo-spatial relations.

This conclusion implies that there are no monadic properties, nor could there be. By "monadic properties" I refer to what are sometimes called "intrinsic" properties, which are not defined or determined in relation to anything different from them. These properties are impossible, since existence is a relation, and so are all the properties that exist in nature.

Everything in nature is hence defined by its temporo-spatial relations to other things that comprise nature. Nature itself is a unity, which is constituted by its internal temporo-spatial relations. To say that something exists is to say that it is part of this complex temporo-spatial structure.

One may wonder, between what entities do these relations obtain? The answer is simple. These relations are all that exists, for there are no other entities in nature other than relations. This may seem problematic. However, a mathematical analogy may help to explain how this is possible. Take, for example, the mathematical eq.

$1 + 1 = 2$. This proposition is a conceptual truth about pure relations, which includes no monadic properties. It describes the relation that obtains between relations.

The reason why mathematics can be used to describe nature with the help of equations, which include nothing but pure relations, is that nature, as I have argued, is comprised of nothing but relations. Furthermore, in Chap. 6 I showed how the conceptual relations that constitute nature provide the units of measurement of time and space, which determine their mutual relation. This is the basis for the use of mathematics for describing nature. It is therefore not a coincidence that mathematics allows describing nature, nor that a mathematical equation can be used as an analogy for understanding the most fundamental relations of nature, that is, the relations between time and space. This is the most fundamental level of existence, and there is no hidden layer supporting it.

The idea of relations that obtain between themselves may seem paradoxical. However, this is not the case. Understanding the relations between time and space is important for resolving the seeming paradox. As the following analysis will show, temporal and spatial relations are complementary, and together create a unity. The relations between time and space, which constitute this structure, are described by the laws of nature. I now turn my attention to understanding the meaning of these laws.

7.3 The Meaning of the Laws of Nature

I have dedicated Chap. 5 to uncovering the structure of the laws of nature. In Chap. 6, I demonstrated how the laws of nature constitute the structure of time and space. Nevertheless, it might justifiably be felt that the meaning of the laws of nature remained unexplained. Although I have shown that they are necessary for constructing an objective conception of nature, and how they constitute both time and space, I have yet to explain their significance in constituting the very essence of existence. In this sense, the laws of nature may seem arbitrary. The question is not how the laws of nature explain the structure of time and space. Rather, the question is, why are there space and time at all?

I shall try to address this question via the notion of change. This is because the laws of nature describe first and foremost how things change, or develop, in time. Although I have explicated the idea of change in temporo-spatial terms, the question nevertheless was left open: Why do changes occur in nature?

Let me try to explain the question I am asking now. I have shown that changes in nature reflect the way different entities in nature determine one another. The changes in nature describe their reciprocal determination. However, it is unclear why different entities should determine one another in time. Part of the answer was given by identifying the entities in nature as determining factors. However, why should these entities be determining factors, and why are they reciprocally determined according to the laws of nature?

An important feature of change, which must be kept in mind while attempting to understand this fundamental aspect of nature, is that change depends on a relation to something else in nature. Recall that my analysis has shown that describing what exists with the help of the distinctions between past, present, and future reflects a relation to a conceptual scheme, which itself is part of nature, from whose point of view what exists is described. Time, as well as change, is therefore relational. As mentioned previously, this relation also constitutes existence.

The notion of time was shown to assume objectivity—a separation between the description and what it describes. This distinction raised the problem of accounting for the connection between the conceptual scheme and what it describes. The solution to this difficulty was given by the laws of nature, which are constituted by the change in spatial relations. This change hence connects the existents in nature, thus constituting the unity of nature. Notice in this context that the laws of nature enable both making the separation between the different elements in nature, in this case, the distinction between the description and what it describes, and the connection between them, thus explaining how what exists can be described, that is, how a connection between two separate entities is possible.

I have argued that existence is nothing but a relation, and that nature is constituted by a complex structure of relations. Exploring these relations led me to acknowledge the existence of the conceptual scheme, from whose point of view nature is described. However, my analysis also led me to the conclusion that the conceptual scheme is part of nature, and itself constituted by temporo-spatial relations. Hence, as far as the exploration of the fundamental features of nature is concerned, the existence of the conceptual scheme plays no fundamental role. However, this does not prevent me from using the insights that I have acquired by investigating the relation between the conceptual scheme and what it describes to shed light on the general relations that obtain between different entities in nature. On the contrary, the fact that the conceptual scheme is like any other entity in nature allows me to rely on the insights that are gained from understanding the relations between the conceptual scheme and what it describes to understand the relations between existents in nature in general.

All these reminders offer the following understanding of change and its role in nature: The distinction between the propositions and what they describe is fundamental to existence. Were it not for this distinction, that is, if there were no way to make the distinction between the proposition and what it describes, there would be no room for making descriptions. There must therefore be a discrepancy between the conceptual scheme and the world. On the other hand, there must also be some type of correlation, otherwise it would be impossible to describe what exists. What explains the ability to maintain the gap between the two while keeping some type of agreement, which allows formulating descriptions of what exists, is the laws of nature. The laws of nature connect all existents in nature, amongst them propositions and what they describe while keeping them separated in time and space.

I wish to emphasize the importance of this distinction, between the conceptual scheme and what it describes, by contemplating the idea that is sometimes described by the term "god's eye view," or by Kant's idea of "intellectual intuition." What is common to such ideas is the suggestion of a possibility of knowledge of things in themselves, without any relation to the subject. However, this view contradicts the conception of nature that I have developed here. To begin with, nature is comprised of nothing but relations, including a relation to the conceptual scheme from whose point of view it is described. There are no monadic properties that can be accessed via supernatural ability. Furthermore, the suggested unity between the subject and the object, as implied by these views, eliminates the distinction between the description and what is described, thus undermining the conceptual structure that gave meaning to the notion of existence. Without the possibility of mistake and ignorance, there would be no distinction between the conceptual scheme and nature, and hence there would be no meaning attached to the idea of describing nature, and the notion of existence.

Recall in this context that the idea of the laws of nature was introduced to close the gap between descriptions of the past, made from the point of view that is located in the present, and what they describe in the past. This distinction was the implication of the notion of objectivity, which was introduced to explain the use of temporal indexical expressions for describing nature—a necessary feature of existence. The laws of nature connect the past and the future, thus allowing us to describe the past. Later, this connection was also found relevant for explaining the connection between simultaneous existents, following the conclusion that simultaneous laws of nature are impossible. Thus, the ability of the conceptual scheme to describe present occurrences was also found in need of explanation. Again, this explanation was provided by the laws of nature, which connect simultaneous existents by their reciprocal determination, that is, the way that they determine one another in the future. Thus, the laws of nature are the basis for the connection between the conceptual scheme and nature.

However, the laws of nature also separate existents. It is important to recognize in this context the connection between time and space. While the laws of nature rely on time to allow existents to determine one another, the distances in time were shown to be founded on distances in space. Thus, space allows making the distinctions between simultaneous existents, whose spatial positions reflect their relations of determination in time. Recall in this context the conclusion that two existents cannot share the same positions in space, because this would imply that they determine one another immediately, which implies that they are the same. In terms of the relation between the conceptual scheme and what it describes, it would mean the collapse of the distinction between the description and what it describes. Space thus allows making the distinction between existents.

A perfect correlation between the conceptual scheme and the world is impossible, as well as between any two existents in nature. A perfect correlation implies the elimination of the distinction between the two entities. While space has the function of separating existents, time enables reconnecting them. This is the role of the laws of nature, which both separate (in space) and connect (in time) occurrences in nature.

The change that is described by the laws of nature allows for closing the gap with time, by creating a correspondence between the conceptual scheme and the world, or more generally between different elements of nature. Thus, this effect can be described as a delayed correlation between different parts of nature.

It is therefore an essential part of nature that there are discrepancies between the elements of nature, together with a delayed correspondence, which is conveyed by the reciprocal determination of these elements. If there were no differences between these elements, there would be no change. In fact, in this case, there would be no distinctions to be made between parts of nature. As already mentioned, eliminating the distinction between two existents means the annihilation of one of them (or both), while eliminating the distinctions between all parts of nature implies annihilation of nature itself.

On the other hand, if there were no way to connect the different elements of nature, there would be no meaning given to the idea of something existing from the point of view of a conceptual scheme. Again, more generally, it can be said that, without any connection between existents, it is meaningless to speak of the existence of nature, that is, of a unified existence.

Although it cannot be said that the change in nature, or the laws of nature, have the purpose of making the elements of nature correspond to one another, this must be the effect of the laws of nature. However, the laws of nature also enable making the separation between the different elements in nature. It is a separation that relies on a delayed reciprocal determination. This allows both the separation and the connection between the different elements of nature, thus constituting nature itself.

As a final note on this subject, notice that this conclusion implies that there can be no complete correspondence between the different elements of nature, and more specifically between nature and the conceptual scheme from whose point of view it is described. Hence, it is impossible to formulate a complete and accurate description of nature. In the context of the conceptual scheme, perhaps the term "information" would be appropriate for formulating this conclusion, by the claim that a conceptual scheme cannot have perfect information about nature, as a matter of conceptual necessity. This implies, not only that a complete description of nature is impossible, but also that a completely accurate description of part of nature is impossible.

7.4 The Structure of Nature

The previous section brought to light an important insight into the fundamental structure of nature. So far, I have discussed existence in nature without addressing the question of the existence of nature itself. In light of the conclusion of the previous section, which explains the meaning of the complementary relation between time and space, I wish to explore this temporo-spatial structure by studying its boundary conditions. In effect, what I am trying to do can be likened to an attempt to annihilate and recreate nature from nothing, thus achieving a better understanding of its constitution.

Before beginning the inquiry, I should clarify that by "creation" I do not mean a process of creation in time. That is, the following discussion should not be understood in temporal terms. The following thought experiment is meant to explore the conceptual structure that was developed in the previous chapters, focusing on the connection between time and space, as mediated by the laws of nature. It is not meant to answer the question of whether nature had a beginning in time, and if so, how was it created. These questions will be discussed in the following section.

I have argued (in Sect. 6.9) that there must be a limit to the speed of determination in nature. In Sect. 7.3 the significance of this fact was revealed. The laws of nature allow making distinctions between different elements in nature while maintaining a connection between them, which is demonstrated in their reciprocal determination. However, the process of reciprocal determination is restrained by the speed of determination. This fact prevents all the different elements of nature from determining each other immediately, thereby eliminating all the distinctions between them. Immediate reciprocal determination hence amounts to mutual annihilation.

Hence, it might seem that a simple method for conceiving the annihilating of nature would be to accelerate the speed of determination. Prima facie, this would decrease distances in space, because space is measured according to the speed of determination, as was demonstrated in Sect. 6.9. Thus, one might conclude that all distances in nature will be gradually reduced as the speed of determination is elevated, and as the speed of determination approaches infinity, all distances in space will approach zero. Due to the connection between time and space, this will also affect distances in time which will also approach zero. Thus, nature as a whole, that is, as a structure of temporo-spatial relations, will be reduced to nothing.

However, this suggestion will not do, as it is based on a misunderstanding of the nature of time and space and their interrelations. The speed of determination reflects nothing but the relations between time and space. It would have been possible to accelerate the speed of determination if time, space, or both, had been absolute entities. However, there is no absolute time that flows at a certain speed, which can be speeded up or slowed down, nor can distances in space be reduced. Since the speed of determination itself is nothing but a relation between time and space, it cannot be changed.

Another way of showing this would be a reductio ad absurdum, that is, to presume that there is an absolute time, and that the "flow of time," as an absolute change, is accelerated. However, this would not be noticed from a point of view that is internal to nature, for example, from the point of view of the conceptual scheme. This is because its own working would be, according to this premise, hastened. Hence, the speed of determination would seem unaffected from the point of view of a conceptual scheme, as well as all the distances in time and space, in contradiction to the initial premise. This again demonstrates the fact that all that there is in nature are relations between its constituents. The speed of determination is only a relation between the basic elements of nature, that is, between time and space, and hence cannot be accelerated or decelerated. Finally, the determinations in past, present, and future are relational and determined in relation to the conceptual scheme.

Hence, the change in these determinations, that is, the "flow of time," is itself relational.

Another misunderstanding, that it is important to be aware of before beginning this inquiry, is that I do not imply, by the claim that the existence of nature is due to delayed reciprocal determination, that the future of nature is necessarily self-annihilation, that is, a slow process by which the different elements in nature come to correspond to one another. Although reciprocal determination can be described as a delayed determination, that is, a process by which the different elements of nature can come to reciprocal correspondence, I have not presented any consideration that would show that this is necessarily so. There is nothing to disqualify, at the current stage of my inquiry, the possibility that nature is in a state of infinite self-determination, in which the different elements of nature continually determine one another—a state of eternal fluctuation—without ever reaching a state of absolute correspondence.

Nevertheless, it is possible, for a better understanding of the connections between the fundamental constituents of nature, to imagine that the speed of determination is immediate. That is, to stipulate that everything in nature determines everything else instantaneously. This stipulation annihilates all the distances in time, which are constituted by the delayed determination in time, which can be imagined to be collapsing to zero. This brings any temporal distance to zero as well, and time can be imagined in this context as passing infinitely fast. Thus, the temporo-spatial relations collapse respectively, and nature is annihilated.

It is helpful to think about this suggestion in terms of the conceptual scheme's ability to describe nature. Recall that I have argued that time reflects the conceptual scheme's limited point of view of nature. This is reflected in the gap between the conceptual scheme and what it describes, which is displayed as distance in space. The laws of nature connect all things in nature. However, this is a delayed connection that maintains the gap between the description and what is described—descriptions of the past are made from a point of view in the present and descriptions of the present are delayed. If it is imagined that the speed of determination is immediate, the temporo-spatial gap between the conceptual scheme and what it describes is lost. Allegedly, the conceptual scheme would have an immediate description of the totality of nature. As opposed to the actual case in nature, in which things determine one another with a delay that prevents correlation, in this hypothetical scenario all things in nature will be perfectly in agreement with one another, and there will be no more room for change. However, this complete correspondence would imply the collapse of all the distinctions in nature, and hence its annihilation.

In light of this insight, it is possible to comprehend how nature can be "created," by imagining a reverse thought experiment, which begins from non-existence, and attempts to reverse the process I have described above. That is, I suggest trying to constitute nature by creating entities that are separated in time and space.

I should make clear that I am not attempting to describe a process that occurs in time, but rather am attempting to explore the idea of such a separation. That is, I do not describe how nature came to be by stipulating a past moment of creation. As I have previously explained, time involves relations between elements in nature, and hence nature cannot be created by a temporal process. On the contrary, the process I am considering is not a process that involves time at all. This is merely a thought experiment, which is meant to explore the conceptual connections between the key elements that constitute nature.

Starting from nothing makes sense, not only because there is "nothing to explain" in this state, but also because it is a state of complete agreement—there are no distinctions to be made, and hence nothing can be identified or determined to exist. Next, it is possible to imagine this state of non-existence as comprised of two types of entities that cancel one another. That is, this state is equal to a state in which two entities determine one another directly, without any delay, thereby canceling one another.

An analogy of a mathematical equation is useful in this context to exemplify the structure of nature. This is, firstly, because it is comprised of pure relations, just as is nature itself. Secondly, the analogy is helpful because the mathematical relations apply to nature—a necessary feature of nature, as demonstrated in Sects. 6.4 and 6.9, in which it is shown how time and space can be measured.

Take for example the equation $(+1) + (-1) = 0$. The right side of this equation, that is, zero, represents non-existence. This equals two relations that, when added up, cancel one another, that is, $+1$ and -1. Although they amount to zero, these relations are separated on the left side of this equation by their location in space. I should add that this equation is a conceptual truth, that is, it is not a contingent proposition, and hence does not describe an existing state of affairs. Its truth is determined by the conceptual relations between its constituents, and hence it can be said to be "empty," or "senseless" in Wittgenstein's terminology. This reminder will be found important in understanding the existence of nature itself, which will be discussed in Sect. 7.6. In the present context, the fact that it is a conceptual truth is important because this equation requires no further explanation—it describes "nothing."

The reader may wonder, what are the temporo-spatial equivalents of the mathematical relations $+1$ and -1 in this analogy? These elements in nature must be such that if they determine one another immediately, they thereby cancel one another (as implied by the analysis of Sect. 7.3). Furthermore, they should be constituted by temporo-spatial relations, which were found to be the necessary features of existence. The temporal relations are basic and simple—simultaneity and precedence. Furthermore, the structure of time was found to be implied by the necessary conditions for supporting contingent propositions, and hence there seems to be no room for flexibility in this respect. The spatial relations, on the other hand, beyond the fact that they must be analogical to the temporal relations (because they are constituted by temporal relations), allow for greater freedom and versatility. Hence, it is mainly the spatial structure that should account for the determining relations which constitute existents in nature. However, I shall not try to develop such an account within the confines of this investigation. I aim to achieve a better understanding of the structure of nature.

I have suggested beginning from non-existence, and thinking of this state as if it is comprised of two entities, which mutually annihilate when determining one another immediately. To prevent such an annihilation, it is necessary to separate these entities, and the only way to do so is by separating them in space, which also implies separation in time, because spatial relations are constituted by temporal relations. This is because separation in space is constituted by delayed determination in time. Thus, imagining the creations of the spatial dimension thereby creates a temporal dimension, in which these entities determine one another.

I have suggested thinking about the initial state of non-existence as comprised of two entities that cancel each other. Although this is helpful for understanding the relations between existents in nature, it is only a means, rather than an accurate description of this initial state. A more accurate way to understand this thought experiment would be not as two preexisting entities that are separated by time and space, but the creation of two entities. The separation in time and space constitutes these entities, which are themselves nothing but temporo-spatial relations.

Thus, it is possible to visualize this creation as beginning with nothing, and stretching a line in opposing directions, which represents the created entities from an imaginary middle point. Thus, a one-dimensional space is created. This dimension, as explained before, is nothing more than a set of possible locations, each representing a relation of delayed reciprocal determination. The space between the two entities is empty. Hence, the distance between the two entities reflects the time it takes them to determine one another. Their existence, therefore, is nothing but a delayed reciprocal determination with other entities in this space, while the space itself is nothing but a mapping of these relations. Since space reflects a *delayed* reciprocal determination, the introduction of space goes together with the introduction of change, which implies time. Notice that time is not a separate element that needs to be introduced into this structure. It is implied by the very notion of space. Thus, the constituents of this simple structure are mutually connected conceptually. The existence of these entities is constituted by their separation in space, which implies connection in time. The existence of this simple system is constituted of nothing but temporo-spatial relations.

The laws that determine even such a simple structure are yet to be determined. One possibility is that these entities annihilate one another, and this structure collapses into nothing by the time it takes the two entities to determine one another. That is, it is possible that the two entities simply come closer in space until they cancel each other. This possibility suggests seeing the creation of the two entities as a closed system, in which the two entities are created by a separation in space, and the loop is closed in time, by the mutual annihilation of the two entities.

This, however, is only one suggestion. For example, it might be that this system is oscillating in time endlessly. That is, it might be that while each entity determines the other, because of the delayed determination, this system never reaches a point of complete correspondence, that is, annihilation.

I have presented an allegedly possible simple system. This was meant to clarify the basic principles of the conception of nature I have developed in the previous chapters. However, determining which system is possible requires further research into the temporo-spatial structure. First and foremost, it is not clear, for example, that a one-dimensional space is indeed possible, and there may be a need for further dimensions, to give meaning to the idea of the flow of time. As mentioned earlier, my end in this thought experiment is only to lay bare the fundamental principles of the internal relationalism conception of nature that was developed in the previous chapters, by demonstrating the connection between time and space.

However, even if such a simple system is possible, it is not representative of nature as we know it. If there are multiple possible structures possible, then understanding the structure of the actual world would have to consider the existence of conscious entities, which are complex enough to realize a conceptual scheme that allows describing nature. This, as I have shown, requires the ability to describe nature from a changing point of view in time. Furthermore, this requires the distinction between nature and its description, and this distinction must be acknowledged in the conceptual structure of the conceptual scheme. That is, the conceptual structure of the conceptual scheme must include an apparatus that enables it to express, for example, the proposition "P(Nφ) ∧ P('¬T' 'Nφ')." This implies, inter alia, memory, that is, records of the past, as well as the ability to make inferences, and so on. It seems impossible to realize this complex structure in a structure that includes only one-dimensional space. However, exploring this question extends beyond the scope of this investigation.

7.5 The Flow, the Arrow, and the Boundaries of Time

Time is the most fundamental aspect of nature, and based on its analysis I have developed the internal relationalism conception of nature. However, there are still unanswered questions regarding time, which I wish to address in this section. The most important question regarding time, which was not addressed in previous discussions, is the asymmetry between the past and the future. Recall that this asymmetry was recognized at the early stages of my analysis as a necessary feature of nature, and in the course of my inquiry was found to be fundamental for supporting the conceptual structure of existence. I have argued that nature must be experienced as advancing from the past to the future; hence, the past must be more accessible than the future for description from a point of view that is located in the present. Without this asymmetry, the entire conceptual structure, which was constructed to explain the need for contingent propositions for describing what exists, collapses.

The asymmetry in time has two different aspects. Although I do not presume that these aspects are independent of one another, it is important to begin the inquiry by carefully distinguishing between them. One is the "flow of time," a term which I have used in this book to describe the change in the determinations of past, present, and future. These are attributed to the occurrences in time, such that an occurrence is first in the future, then present, and finally recedes into the past. The second is the arrow of time, which is related to the asymmetry in occurrences in time, including in the conceptual scheme. This asymmetry is reflected in different ways. In my analysis, this asymmetry was found, first and foremost, to be manifested in the fact that there are non-inferential descriptions of the past, but not of the future—that is, that we remember the past but not the future. Such asymmetry gives time a "direction," in contrast to the dimensions of space.

To begin with the change in the determinations of past, present, and future, it should be noted that the question of the flow of time has been a subject of a long debate in philosophy. In this debate, the dominant views are represented by two rival theories of time, the A-theory and the B-theory (see Zimmerman, (2005)). These theories were developed as a reaction to McTaggart's argument for the unreality of time (McTaggart, 1908, 1927).

According to the A-theory of time, which is sometimes termed "the tensed view," past, present, and future are real properties of nature, and the change in these properties is an objective feature of nature, which has no relation to a perceiving subject (see, for example, Markosian, 1993; Prior, 1967, 1968). According to this view, it is the change in these properties that constitutes time as the dimension of change (in contrast to space). According to the B-theory of time, which is sometimes referred to as "the tenseless view," on the other hand, these properties are merely subjective, and hence reflect nothing beyond a relation to the subject (Mellor, 1998; Oaklander & Smith, 1994; Smart, 1963, 2008; Williams, 1951). That is, the claim that something is in the past should be translated as the claim that it is earlier to the subject or to a specific attribute of the subject (for example, to making this claim); while the claim that something is in the present is analyzed as a claim that it is simultaneous to the subject or a specific attribute of the subject. Hence, according to this view, the "flow of time" is an illusion, or rather a matter of perspective, rather than a true feature of nature.

Notice that both these views rely on a realist assumption, according to which nature does not depend on the point of view from which it is described. Hence, the proponents of the B-theory of time, since they find the distinctions between past, present, and future to be subjective, are led to deny that the change in these distinctions constitutes time. Proponents of the A-theory of time, on the other hand, because they find the distinctions between past, present, and future necessary for constituting time, are led to deny that these distinctions bear any relation to the perceiving subject.

The conception of time I have developed here fundamentally differs from both the A-theory and the B-theory of time, the two prevailing opposing views on the nature of time. It should be noted that both theories are formulated against the backdrop of our actual conceptual scheme and therefore rely on complex conceptual apparatuses and assumptions that my analysis seeks to avoid. More specifically, despite superficial similarities to the A-theory of time, my analysis does not share its semantic, ontological, or epistemological premises. Thus, common criticisms of the A-theory, and specifically its attempts to establish temporal relations based on distinctions between past, present, and future, do not apply to my analysis (see, for example, Boccardi, 2015, 2018; Oaklander, 1994).These criticisms are usually directed against specific versions of the A-theory, which include substantial ontological commitments or rest on a commonly held assumption that I do not share—namely, that the distinctions between past, present, and future are empirical and hence based on experience.

Furthermore, it may seem that my analysis, due to its abstract nature, is detached from the lived experience of temporal passage, a central concern in much of contemporary philosophy of time. However, I view this as an advantage rather than a shortcoming. First, my analysis attempts to reconstruct temporal distinctions from nothing—that is, without any unjustified premise about nature. Moreover, the very concept of "experience" is complex, as it relies on an intricate conceptual structure that includes assumptions about nature, subjects of consciousness, and their mutual relations. Furthermore, the conclusions of my analysis show that the structure of time is necessary rather than contingent; hence, it cannot be studied by exploring our experience, which, as mentioned earlier, can describe only how things are, not how they necessarily must be.Internal relationalism denies the claim that nature exists in an absolute sense, without any relation to another entity. Although the existence of nature is not relative to any entity outside of it, the existence of nature is determined in relation to entities that are internal to it. This idea will be further discussed in the next section. In the present context, what is important is the conclusion that existence is a relation, for example, between the conceptual scheme and the nature of which it is part. Furthermore, according to my analysis, the fundamental aspect that explains existence is the change in the distinction between past, present, and future, which are used to describe nature from the changing point of view of the conceptual scheme. Hence, although according to this view, the distinctions between past, present, and future are subjective, as maintained by the B-theory of time, they are nevertheless constitutive of time, as maintained by the A-theory of time.

Finally, an important aspect of the controversy between the A-theory of time and the B-theory of time is the ontological status of the occurrences in the past, present, and future. The B-theory is forced to reject any ontological distinctions between past, present, and future occurrences, because of its view of these distinctions as subjective. Hence, according to this view, all occurrences, in the past, present, and future, equally exist simpliciter—a view that is termed "eternalism" (see, for example, Sider (2001); Smart (1963); Taylor (1955);. Williams (1951)). The A-theory, on the other hand, can accept eternalism. This is because the distinctions between past, present, and future, even if considered objective, need not have an ontological implication—a "moving spotlight" version of the A-theory (see, for example, Cameron (2015); Deasy (2015)). However, the A-theory can give the present a unique status, according to which only present occurrences exist simpliciter—a view that is termed "presentism" (see, for example, Bigelow (1996); Bourne (2006); Markosian (2004); Prior (1972); Zimmerman (2007)). Finally, the A-theory allows a distinction between the past and the present, on the one hand, and the future, on the other, and accepts the view that only past and present occurrences exist—the "growing block theory" (see, for example, Broad (1923); Tooley (1997)).

The conception of nature that I have developed here explains existence as a relation. However, it is not simply a 2-place relation. Rather, existence in nature is determined in relation to a complex structure of temporo-spatial relations. Hence, although according to this conception of nature the "flow of time" is constitutive of existence, similar to the A-theory of time, it is forced to reject the ontological distinctions between past, present, and future. Furthermore, it is forced to reject the distinction between existence-in-the-past, existence-in-the-present, and existence-in-the-future, on the one hand, and "existence-simpliciter," on the other, on which this debate is founded.

The change in the temporal determination of past, present, and future, however, is a necessary feature of nature. As such, not only is it not itself an occurrence, but it is also not something that can be experienced (Yehezkel, 2013). Furthermore, it is difficult to see how the flow of time enables explaining the asymmetry in occurrences in nature, that is, the arrow of time. How can the flow of time explain, for example, the fact that we have a non-inferential description of the past, that is, the memory of the past, but do not have non-inferential descriptions of the future, that is, the memory of the future? How can the flow of time explain the fact that occurrences in nature seem irreversible, for example, the fall of a glass cup on the floor and its breaking into little pieces?

Thus, the question of the flow of time is prima facie distinguished from the question of the arrow of time, which is related to temporal asymmetry in occurrences in nature. Although this asymmetry in phenomena in time is also a necessary feature of nature, in contrast to the flow of time, it is not clear what constitutes this aspect of time. This question has been discussed extensively in the philosophical literature (see, for example, Newton-Smith (1980); Price (1996); Savitt (1995)). However, in the confines of this book, I limit the discussion to the question of how the arrow of time can be explained in the terms of the conception of nature that I have developed in the previous chapters.

The difficulty of explaining the asymmetry that is reflected in the arrow of time, within the boundaries of the internal relationalism conception of nature seems even graver, in light of the conclusion of the analysis of the structure of the laws of nature, which describe the occurrences in nature. Recall that I have argued that the laws of nature, which connect occurrences at different times, are symmetrical, in the sense that the occurrences they connect in different times are necessary and sufficient for one another. In other words, there is no asymmetry in the structure of the laws of nature between later and earlier times, such that could explain the asymmetry of the arrow of time.

Furthermore, in Sect. 7.4 I have suggested a simple temporo-spatial structure, to demonstrate how nature can be created out of nothing. I have tried to demonstrate, using examples, the connections that obtain between time and space, which ultimately constitute nature. Thus, it was shown that the creation of existents is given meaning by spatial relations, which themselves imply temporal relations. Notice that these mutual relations between the fundamental features of nature imply that any such description can be reversed. Creating or annihilating existents implies separating or combining these entities in space and time, while time and space relations are mutually defined. The relations between the fundamental aspects of nature are therefore reversible. Hence, for example, the simplified system I have described in the previous section is reversible in time. There is nothing in what I have described there that can give meaning to the idea of the arrow of time. This, however, only shows that this system is far too simplistic to explain the arrow of time.

Explaining the arrow of time hence requires a more complicated structure than that which is discussed in the previous section. To give meaning to the asymmetry in occurrences in nature, this asymmetry must be reflected in the occurrences themselves.

The discussion on the arrow of time may seem to lead me away from conceptual analysis and into the contingent structure of nature. Not only does it appear that any discussion on this issue requires an analysis of the physical processes in nature, such as the breaking of a glass and the working of the memory, but it also might be thought that explanation of the arrow of time can only be given by describing the workings of complex physical processes. Indeed, the question of the arrow of time is extensively discussed in modern physics, and there are many suggestions for explaining the arrow of time, based on current physical theories (Savitt, 1995).

My investigation in this book is conceptual, and hence discussions of contingent aspects of nature extend beyond the scope of my discussion. Hence, although I do not believe that it is possible to offer a purely contingent explanation for the arrow of time—such suggestions may explain some of the phenomena that are associated with the arrow of time, but I believe that they presume a metaphysical asymmetry—I shall not critically discuss such attempts. My main goal in the current discussion is to show how the conceptual structure that I have developed here offers a solution to the problem of the arrow of time, while suggesting how the structure of space can be relevant for explaining this asymmetry in time. This is important, because this asymmetry is a necessary feature of nature, according to my conception of existence, and hence the solution for this riddle must be found within the confines of the conceptual structure that I have developed in this book.

In order not to beg the question, I have begun the discussion with the distinction between the flow of time and the arrow of time. However, I believe that the key to explaining the arrow of time, that is, the asymmetry found in occurrences in time, lies in the flow of time, that is, the change in the determination of past, present, and future. I should emphasize, however, that I do not claim that this change in the determination of past, present, and future has ontological implications, in the sense, for example, that the past or the future do not exist. To begin with, ontological distinctions between the past, present, and future do not make sense in the confines of the conception of time I developed here, in which existence is interpreted as a relative term. Furthermore, it should be noted that ontological distinctions of this type cannot explain the arrow of time, which is reflected in the structure of occurrences in time. Nevertheless, this change does reflect an asymmetry in time, which relates to how earlier occurrences determine later occurrences. I have argued that earlier and later occurrences, which are connected by the laws of nature, are both necessary and sufficient conditions for one another. Hence, there is a sense in which it is legitimate to say that later occurrences determine earlier occurrences. This is in the sense that if one knows later occurrences, one can infer the previous occurrences.

Nevertheless, according to the conception of nature I have developed in the previous chapters there is a fundamental difference between earlier and later occurrences. My analysis shows that the connection that is established by the laws of nature, between occurrences at different times, is that of a delayed determination, and this delayed determination implies that previous occurrences have a priority over later occurrences, which is demonstrated in an asymmetry in time. Although I do not intend to give a detailed account of the concept of causality, I believe that this is the reason behind the asymmetry in this concept. That is, this is the reason we think of past occurrences as influencing later occurrences, rather than vice versa.

However, even if this asymmetry is accepted, it is not clear how, given the asymmetry in the laws of nature, this metaphysical symmetry can be reflected in occurrences in nature. The solution for this difficulty, I believe, is found in the structure of space. Recall that I have explained the structure of space as echoing the relations of reciprocal determination between simultaneous entities. It is therefore expected that this asymmetry in time will be exhibited in the way in which occurrences influence one another in space.

Thus, if an occurrence φ at time T is determining later occurrences, this determination is expanded in time, throughout space, from the location of φ at the speed of determination (see Sect. 6.9). Suppose that something happens at a certain location in space, for example, a glass cup is tipped over a table. This determines its surrounding occurrences, for example, the movement of particles of air near the falling glass, which in turn determines a change in more remote particles in the surrounding area. Later the glass hits the floor and sends shock waves through the floor in expanding circles, and the breaking of the glass spreads fragments of broken glass, expanding through space.

Although the process of the falling and breaking of the glass cup can be reversed without breaking the laws of nature—for they merely describe the interchangeable relations between temporo-spatial relations—there is a clear direction to this process of reciprocal determination. If it is imagined as reversed, the coordinated movement of the parts of that glass, as well as the shock waves passing through the floor, as they come together to create the glass and push it up to the table, seems unnatural. This is because it seems that instead of one occurrence determining the future occurrences in its adjacent space, the surrounding occurrences join forces in determining one occurrence. This is why, given the two occurrences in time, it is natural to conclude that the fall of the glass cup caused it to break and spread broken glass all over the kitchen floor, rather than vice versa.

The direction of time, therefore, is expressed in spatial terms, that is, it is expressed in the way one occurrence's determination of its neighboring occurrences expands with time in space. This is an example of the interconnection between time and space, and the importance of the structure of space. For such a direction cannot be found, so it seems, in the simplified system I described in Sect. 7.4, which includes only one spatial dimension. This does not explain why space is three-dimensional—prima facie it seems that two-dimensional space is sufficient for giving meaning to the arrow of time. However, it does show how the structure of space is significant for shaping occurrences and giving meaning to the arrow of time. This should not come as a surprise, given the close connection between time and space. Recall in this context that space is constituted by temporal relations, and was introduced to give substance to the idea of a change in time.

The fact that positions in space represent a relation of delayed determination of neighboring occurrences also explains the asymmetry between non-inferential descriptions of the past and the present. The state of the conceptual scheme is influenced by previous occurrences, and the previous states of the conceptual scheme influence its later states, thus enabling memory and making inferences in time. All these allow the conceptual scheme to expand its description of nature in time. This also explains how the arrow of time is identified. However, it does not imply that the arrow of time is not merely a subjective phenomenon. All these effects are the results of a more basic asymmetry in time, which is explained by how past occurrences determine, or cause, later occurrences.

My discussion of the asymmetry of time naturally leads me to the question of the boundaries of time, that is, the question of whether time has a beginning or an end. These questions have been debated in vain for millennia, with both options raising difficulties and seeming unacceptable to many. However, they were usually conducted under a realist conception of nature. Kant, on the other hand, in light of his conception of nature, undermined both options and claimed that they rely on the false assumption that nature (or "world) is a thing in itself (Kant, 1929, p. B533). However, Kant's conception of transcendental idealism is fundamentally different from the internal relationalism conception of nature that I have advanced here. The question I wish to address is whether there is a definite answer to this question according to the internal relationalism conception of nature.

The relevance of the asymmetry in time to the question of the boundaries of time is obvious. Usually, discussion of the boundaries of time revolves around the question of whether nature has a beginning, rather than whether nature has an end. Both common options, that is, that nature has a beginning in time and that nature exists for an infinite time seem problematic. However, the question of whether nature will end or exist forever does not seem to raise the same difficulty. Kant, for example, formulated his first Antinomy only concerning the question of the earlier boundary of nature in time, that is, the question of whether it has a beginning or not (Kant, 1929, p. B454).

The previous analysis justifies this asymmetry. Time is indeed asymmetric, and although earlier and later occurrences are necessary and sufficient conditions for one another, it is the earlier occurrences that explain the existence of later occurrences. This is the reason why, while trying to explain the existence of nature, it is necessary to venture into its past, rather than into its future. Hence, I shall not discuss the question of the future of nature. The fact that nature is explained by its past, rather than its future, seems to imply that there are no straightforward a priori considerations that would settle the question of whether nature will continue to exist forever or not. I should emphasize at this point that my previous claim (in Sect. 7.4), that if the speed of determination would be instantaneous then nature would be annihilated, does not imply that nature will be annihilated in the future. As I have explained, the thought experiment in which the speed of determination is thought of as immediate is not equivalent to accelerating the speed of determination, or to speeding the passage of time, and hence does not predict the future of nature. Similarly, the imagining of the creation of nature (in Sect. 7.4) was not meant to be taken as a creation in time, and hence does not imply anything about the origin of nature in time.

The question of whether nature was created in time or not should be discussed in terms of the conception of nature I have developed here. Recall that I have introduced a relational conception of nature. According to this conception of nature, its existence can only be determined in relation to an entity that is part of nature, that is, internal to it. Furthermore, the most fundamental feature of nature, which gives meaning to the idea of existence, is time itself, which is constituted by a relation between the existents in nature. Since this structure is constituted by the laws of nature, which describe the relations that obtain between the fundamental features of nature, the question of the origin of nature is determined by the consistency of the different options with the laws of nature.

I begin with the suggestion that nature existed for eternity. According to this suggestion, not only does nature not have a beginning in time, but the duration of its existence also cannot be measured by a finite number. That is, it existed for an infinite time in the past. The question of whether this is possible is translated as the question: Is it possible to formulate an objective explanation of the current state of nature, while relying on the premise that it existed for unlimited time in the past? The difficulty that arises due to this suggestion is that, as the previous discussion has shown, the present state of nature must be explained by the past states of nature. Thus, each state of nature is explained by the previous state. The assumption that nature existed for infinite time before the present moment hence leads to an infinite regress. That is, this assumption implies it is impossible to formulate a complete description of nature that explains its current state.

Notice that this problem, which within the confines of an absolute conception of existence would merely be an epistemological problem, is inconsistent with the relational conception of nature that I have developed here. Although I have argued that there are limitations to the description of nature, which arise because of the limited point of view of the conceptual scheme, the implication that a complete explanation of nature is impossible is unrelated to this limitation. Recall that the objective description of nature is formulated by the current state of the conceptual scheme together with the laws of nature. That is, according to this conception of nature, the current state of nature is connected by the laws of nature to the point of view from which nature is described. This is not a mere epistemological issue. This constitutes the objective realm, that is, nature. These laws of nature explain how the temporo-spatial relations, which constitute the current state of nature, have been determined by their past states. Since the implication of the premise that nature existed for an infinite time before this moment in time implies that no finite explanation can be given, that is, no complete explanation can be given for the current state of nature, this amounts to the conclusion that no determinate state of nature exists. For it implies that there is no objective description of nature—that there is nothing that determines its current state. The suggestion that nature existed for an infinite time in the past is hence inconsistent with internal relationalism.

It might be thought that the previous consideration, for rejecting the suggestion that nature has existed for infinite time, leaves open the possibility of a cyclic nature, in which occurrences repeat themselves in an infinite sequence of repeating occurrences. This suggestion is more difficult to dismiss. To begin with, this general description leaves open many different models of cyclic histories, and it is difficult to formulate a conclusive argument that can disqualify them all. Discussing these possibilities extends beyond the scope of this book. I shall hence limit the discussion to pointing out what seems to me the biggest problem with this suggestion—the explanation of the arrow of time. Prima facie, a cyclic nature would require occurrences to be completely symmetrical in time. However, this does not cohere with the existence of the arrow of time, which is implied by the flow of time—a necessary feature of nature. Thus, is difficult to see how these ideas, of a cyclic nature and asymmetry in time, can cohere.

It might seem that the possibility of a cyclic nature can be defended by a process of elimination, that is, in light of the shortcomings of the rival models. This is because the only possibility that is seemingly left is that nature has a beginning in time, which prima facie seems even more problematic.

To begin with, the suggestion that nature has a beginning in time seems to violate the laws of nature. Recall that the laws of nature require connecting occurrences at different times, such that these occurrences are necessary and sufficient conditions for one another. Furthermore, these laws are constituted by temporo-spatial relations. Hence, it is impossible to explain, with the help of the laws of nature, how nature came to existence in time from nothing. As previously mentioned, this is unacceptable in the confines of internal relationalism.

In the confines of this discussion, which relates to this question exclusively in terms of the internal relationalism conception of nature, I do not attempt to survey the arguments on the beginning of time (see, for example, Newton-Smith (1980), Chap. 5). However, a relevant consideration is found in the writing of Aristotle. Aristotle has correctly argued that time cannot begin, since there is no earlier or later without time (Aristotle, 1998, p. 1071b). That is, the claim that time began at one point is self-contradictory, since it implies that there was no time before that point. However, if there was a "before," there must have been time. This claim, therefore, implies a contradiction.

Indeed, it is also true, according to internal relationalism, that time cannot begin. For according to this view time is constituted by a relation between parts of nature. It is meaningless to talk of time that is not internal to nature. Hence, the claim that time began at some point in time is meaningless. Time is constituted by the laws of nature, and only these laws identify points in time, and these connect occurrences at different times by necessary and sufficient relations. The claim that there was a point in time that nothing precedes must be based on the laws of nature. However, the laws of nature cannot determine that there was a time in which nature did not exist. This is indeed a self-contradiction.

However, the rejection of the suggestion that nature has a beginning in time does not imply a refutation of the suggestion that nature has a finite history. Two options are seemingly compatible with the idea that nature had a finite history but not a beginning in time. The first is a cyclic nature in which time repeats itself. However, this option again raises the problem of accounting for the asymmetry of time, which is a necessary feature of nature. That is, it seems that this suggestion requires time to run backward at some stage, and this claim is self-contradictory.

The second option, for examining how nature can have no beginning but is nevertheless finite, does not involve a cyclic history. To understand how this is possible, suppose that the more you go back in time, the smaller nature gets in space, and hence the temporal lengths that are correlated with these spatial distances. However, although the more you go back in time, the smaller the temporo-spatial lengths become, there is no point at which nature reaches size zero and annihilates. The most that can be done is to asymptotically approach a hypothetical and unreal point of beginning. Thus, while nature has no beginning in time, its age is finite (for a similar suggestion, see Smith (1985)).

The idea that there is no starting point to nature, but it is nevertheless finite in time, is only problematic from a realist view of nature. In a realist conception, this suggestion would not allow explaining the existence of nature. However, in an internal relationalism conception of nature, it makes perfect sense that it is impossible to reach the limit of nature in the past. Furthermore, this model allows giving a complete account of the current state of nature, for it does not involve an infinite regress in time. It sets a limit to the explanation of nature, without positing a starting point.

These considerations are not conclusive, for I have not disproved the idea of a cyclic nature. However, such a model of nature does not seem to raise the difficulties that are associated with cyclic models of nature. As already mentioned, the biggest problem, in my view, with cyclic models of nature is the inability to account for the asymmetry of time. Recall that, in a previous discussion about the arrow of time, I used the expanding pattern of determination of a single occurrence of its neighboring occurrences in space to explain the asymmetry between earlier and later occurrences. The fact that the model I have described above follows a similar pattern, that is, of a nature that is expanding in space with time from an infinitesimally small point, shows that this model can directly account for the asymmetry of time. Furthermore, the considerations that relate to the asymmetry of time can likewise explain why such a pattern describes nature as a whole.

Although my suggestion does not provide a causal explanation for the arrow of time, this is, of course, all that can be expected from a conceptual analysis. An a priori inquiry into nature cannot offer a physical theory of phenomena. However, I believe that the arrow of time is one area where natural philosophy and empirical science are already closely aligned and can collaborate to formulate a solution to this puzzle. This collaboration, however, goes beyond the scope of this inquiry. The relationship between these two research projects will be discussed further in the book's conclusion.

7.6 Why Is There Something Rather Than Nothing?

The question "why is there something rather than nothing?" is the most profound question about nature. This is a question about existence per se, which is the final end of any attempt to understand nature. A failure to answer this question indicates a fundamental shortcoming of any suggested conception of nature, while an ability to answer this question is its ultimate test.

In light of the discussion of the previous section, I should make clear that the question "why is there something rather than nothing?" is independent of the question of the origin of nature in time. That is, the question of why nature exists arises whether nature has a beginning in time or not. A nature that existed for eternity needs explanation just as does one that was created in time.

Internal relationalism offers a straightforward answer to the question of why nature exists. Recall that, according to this conception of nature, existence is relative. That is, it is impossible to speak of the existence of anything independently of its relations to other entities. Thus, for example, the question of why the sun exists should be answered by its relation to the other entities.

This conclusion implies that any proposition that determines existence, for example, that "φ exists," should be analyzed as describing a relation. Since existence is a relation, if no other entity is mentioned in relation to which φ is determined to exist, it is determined in relation to this description itself (which reflects the point of view of a conceptual scheme).

The explanation for the fact that existence is relative was introduced in Sect. 7.4, in which it was demonstrated how the creation of one entity is related to other entities in space and time. However, I should emphasize that the fact that existence is relative is independent of this explanation. It is a conclusion of the analysis of the concept of existence, and of the necessary conditions for the meaningfulness of contingent propositions, which describe what can be meaningfully said to exist (see Sect. 7.2).

So far, I have discussed the existence of entities in nature. However, the same conclusion equally applies to nature itself. The claim that nature exists is determined in relation to other entities. Since there are no other entities other than the ones which comprise nature, its existence is determined in relation to entities that are part of it. Thus, the claim that nature exists should be analyzed as describing a relation to the point of view from which this claim is made, that is, a point of view that is internal to nature. There is no absolute point of view that is external to nature and independent of it from which it can be meaningfully claimed that nature exists.

Since existence is a relation, the claim that something does not exist should be equally analyzed as relational. That is, something can be said not to exist only in relation to something else. Again, when no other entity is mentioned, in relation to which this claim is made, this assertion should be interpreted as relating to this description itself. Thus, the claim "there is nothing" can only be made in relation to something, and in this case, the proposition itself. The proposition "there is nothing" therefore presumes the existence of a point of view in relation to which there is nothing. Hence, the claim that nothing exists is a self-contradiction.

It follows from my analysis that the proposition "there is something" is a conceptual truth. The meaningfulness of this claim implies that there is something, that is, the conceptual scheme from whose point of view this claim is made. Hence, the answer to the question "why is there something rather than nothing?" is that something necessarily exists. The existence of nature is therefore a necessary truth.

I wish to explain the meaning of my conclusion. Because it is impossible to claim that nature exists in an absolute sense, the only meaningful way to claim its existence is from a point of view internal to nature. However, from a point of view that is internal to nature, nature necessarily exists. Again, nature is simply a complex structure of temporo-spatial relations. This structure is itself a conceptual structure, which is established by the necessary connections that connect its constituents. Nature is therefore a necessary conceptual structure.

This conclusion may seem to undermine the claim that only the things that exist are described by contingent propositions. In light of this conclusion, which was the starting point of my investigation into the nature of existence, how can nature be said to exist necessarily?

Indeed, the claim that nature as a whole exists is not a meaningful statement. It is necessary, and hence, as I suggested before (in Sect. 2.4), should be understood as an empty sentence, that is, devoid of any descriptive meaning. A necessary truth does not describe what exists, but rather reflects the conceptual connections between the concepts that constitute it. The claim that nature necessarily exists demonstrates the limiting condition for the use of the concept of existence. It is only meaningful to ascribe existence in relation to something else, and nature as a whole can only be said to exist in relation to part of it. Hence, when one speaks of the existence of nature, it becomes a conceptual truth, a sign that the limit of the legitimate use of the concept of existence has been reached. Meaningful use of the concept of existence can only refer to a relation between elements of nature, and in this use, it does formulate a contingent proposition.

The conclusion of this section is thus not only consistent with but also coheres and in a sense completes the analysis of the concept of existence, which was undertaken in the course of this book. The conditions for the meaningfulness of the concept of existence were found to be identical to the conditions for the meaningfulness of contingent descriptions. The analysis of contingent propositions found that they reflect the point of view that is internal to nature, from which existence is described. This relation was also found to explain the fundamental features of nature, that is, temporo-spatial relations. These are the relations that constitute (contingent) existence. To say that something exists is to say that it exists in time and space, and hence exists as a part of the temporo-spatial structure of interdependent relations. The existence of this temporo-spatial structure can hence only be determined in relation to something internal to it, and hence its existence is necessary.

References

Aristotle. (1998). *The metaphysics* (H. Lawson-Tancred, Trans.). Penguin Publishing Group.

Bigelow, J. (1996). Presentism and properties. *Philosophical Perspectives, 10*, 35–52.

Boccardi, E. (2015). If it ain?T moving it shall not be moved. *Topoi, 34*(1), 171–185. https://doi.org/10.1007/s11245-013-9230-7

Boccardi, E. (2018). Turning the tables on mctaggart. *Philosophy, 3*, 1–16. https://doi.org/10.1017/s0031819118000141

Bourne, C. (2006). *A future for presentism*. Oxford University Press.

Broad, C. D. (1923). *Scientific thought*. Routledge and Kegan Paul.

Cameron, R. P. (2015). *The moving spotlight: An essay on time and ontology*. Oxford University Press.

Deasy, D. (2015). The moving spotlight theory. *Philosophical Studies, 172*(8), 2073–2089.

Kant, I. (1929). *Critique of pure reason* (N. Kemp-Smith, Trans.). Macmillan.

Markosian, N. (1993). How fast does time pass? *Philosophy and Phenomenological Research, 53*(4), 829–844.

Markosian, N. (2004). A defense of presentism. *Oxford Studies in Metaphysics, 1*, 47–82.

McTaggart, J. M. E. (1908). The unreality of time. *Mind, 17*(68), 457–474.

McTaggart, J. M. E. (1927). *The nature of existence* (Vol. II). Cambridge University Press.

Mellor, H. D. (1998). *Real time II*. Routledge.

Newton-Smith, W. H. (1980). *The structure of time*. Routledge & Kegan Paul.

Oaklander, L. N. (1994). Bigelow, possible worlds and the passage of time. *Analysis, 54*(4), 244–248. https://doi.org/10.2307/3328812

Oaklander, N. L., & Smith, Q. (1994). *The new theory of time*. Yale University Press.

Price, H. (1996). *Time's arrow and archimedes' point: New directions for the physics of time*. Oup Usa.

Prior, A. N. (1967). *Past, present and future*. Clarendon Press.

Prior, A. N. (1968). *Papers on time and tense*. Oxford University Press.

Prior, A. N. (1972). The notion of the present. In J. T. Fraser, F. C. Haber, & G. H. Muller (Eds.), *The study of time* (pp. 320–323). Springer Verlag.

Savitt, S. F. (Ed.). (1995). *Time's arrows today: Recent physical and philosophical work on the direction of time*. Cambridge University Press.

Sider, T. (2001). *Four dimensionalism: An ontology of persistence and time*. Oxford University Press.

Smart, J. J. C. (1963). *Philosophy and scientific realism*. Routledge.

Smart, J. J. C. (2008). The tenseless theory of time. In T. Sider, J. Hawthorne, & D. W. Zimmerman (Eds.), *Contemporary debates in metaphysics* (pp. 226–238). Blackwell.

Smith, Q. (1985). On the beginning of time. *Noûs, 19*(4), 579–584.

Taylor, R. (1955). Spatial and temporal analogies and the concept of identity. *Journal of Philosophy, 52*(22), 599–612.

Tooley, M. (1997). *Time, tense, and causation*. Oxford University Press.

Williams, D. C. (1951). The myth of passage. *Journal of Philosophy, 48*(15), 457–472.

Yehezkel, G. (2013). The illusion of the experience of the passage of time. *Disputatio, 5*(35), 67–80.

Zimmerman, D. W. (2005). The A-theory of time, the B-theory of time, and 'taking tense seriously'. *Dialectica, 59*(4), 401–457.

Zimmerman, D. W. (2007). The privileged present: Defending an "a-theory" of time. In T. Sider, J. Hawthorne, & D. W. Zimmerman (Eds.), *Contemporary debates in metaphysics* (pp. 211–225). Blackwell.

Chapter 8
Conclusion

Abstract In this, the final chapter of this book, I summarize the results of my investigation of nature, and outline directions for future research. I argue that empirical science and natural philosophy should not be seen as competing methodologies, but rather as complementary inquiries into nature. Each has its limitations and advantages, and hence, for a comprehensive understanding of nature, both projects should be pursued.

Keywords Conceptual research · Natural philosophy · Science

In this book, I have attempted to establish a foundation for the conceptual research of nature. I began by developing the conceptual tools that are needed for conceptual analysis (in Chap. 2). These tools were first implemented (in Chap. 3) in an analysis of the concept of existence, which showed that only what is described by a contingent proposition can be meaningfully said to exist. I then attempted (in Chap. 4) to develop the conceptual apparatus that supports the use of contingent propositions, and the conceptual structure that was constructed during this inquiry was identified with the temporal indexical distinctions of past, present, and future. Further analysis of this structure revealed that it relies on the distinction between the propositions and what they describe, which in turn required the introduction of the distinction between the conceptual scheme and nature, of which this conceptual scheme is part. Establishing this distinction required the introduction of the laws of nature (undertaken in Chap. 5), which binds together occurrences at different times, such that these occurrences are necessary and sufficient conditions for one another. Furthermore, these laws were ultimately found (in Chap. 6) to constitute objective temporal relations, as well as the relations between simultaneous occurrences, that is, space.

The conception of nature that I have developed here, which I term internal relationalism, has profound implications (which were explored in Chap. 7). According to this view, existence is a relation between elements of nature, and nature is constituted by temporo-spatial relations, that are established by the laws of nature. This conception of existence explains how things are distinguished by their separation in

G. Yehezkel, *Conceptual Principles of Natural Philosophy*, Synthese Library 495, https://doi.org/10.1007/978-3-031-82108-0_8

space and connected in time by reciprocal determination. This is the most basic framework of relations in nature, upon which all that exists is supported.

This conception of nature explains many aspects of nature as we know it. It explains the structure and significance of time, the structure of the laws of nature and their role in shaping the objective realm, the connection between time and space, the arrow of time, and finally, the nature of existence. However, there are countless questions about nature that are yet to be answered. The test of any pretention for establishing the philosophy of nature as a viable science is in the prospect of expanding the scope of its research, thus advancing the understanding of nature.

The history of established sciences, such as mathematics and physics, indicated two separate aspects in which science can and should advance. To begin with, an important aspect of the progress of any science is methodological. Although I have introduced the conceptual tools and principles that should direct and govern the philosophical inquiry of nature, it is expected that future research will further enrich the tools and techniques of conceptual research, as well as improve the accuracy and the rigorousness of the analysis, which would secure the validity of its conclusions.

The second aspect in which the success of any science is evaluated is its ability to continue providing answers to the questions that are left open in its field of research. The richness of nature leaves numerous questions and puzzles which are yet to be answered. However, in the course of the inquiry undertaken in this book, many lines of inquiry were left open, which I find promising for future research. The structure of the laws of nature can be further developed and explored; the study of space was undertaken in a very rudimentary way, which leaves much to be investigated; the implications of the characterizations of simultaneity and the speed of determination have yet to be studied and explored, and so on.

This leads me to the questions: What can be expected from an a priori research of nature? What are the limits of such an inquiry? And finally, what should be the goal of this endeavor? Answering these questions requires addressing the question of the relation between the philosophy of nature, as a conceptual a priori science, and the empirical science of nature, that is, physics.

Recall that I have argued that the only things that could be meaningfully said to exist are those that are described by contingent propositions. In light of this conclusion, one may wonder, how the a priori investigation of nature is possible, given that it can only describe what is necessary.

I trust that the conceptual investigation, which was undertaken in the previous chapters, has established the feasibility of the conceptual study of nature. Furthermore, it explained its own possibility, as well as the relations between the philosophy of nature and physics.

The analysis of the concept of existence has shown that only the things that are described by contingent propositions can be meaningfully said to exist. This established the existence of the contingent realm, whose exploration can only be undertaken empirically. This is the task of physics, as the most fundamental empirical science of nature. However, it was also shown that the contingency of propositions is explained by the fact that nature is described from a point of view that is located in nature. This point of view is necessarily internal, limited, and changing in time, and these features explain why contingent propositions are needed for describing nature.

The contingent realm can only be explored empirically. Hence the need for an empirical science of nature. However, experience can only teach us that things are, as a matter of fact, so and so. It can never teach us that things are necessarily so. To take a trivial example, looking outside can teach you that it is raining now, and hence the truth of the proposition that "it is raining outside now or it is not raining outside now." It can never teach you that it is a necessary truth.

The fact that there are necessary truths, which go beyond anything that the empirical exploration of nature can offer us has, I trust, been established beyond any shadow of a doubt by the analysis of the previous chapters. The same analysis of the concept of existence, which showed the need for empirical research of nature, has successfully exposed and established the fundamental features of nature, that is, time and space, as well as uncovered the necessary structure of the laws of nature, which are presumed by the empirical sciences. Furthermore, it offered important insights into the nature of space and time and their interconnection, as well as answers to some of the profound questions that have puzzled philosophers and physicists for generations.

Natural philosophy and physics have the same goal, that is, a deep and thorough understanding of nature. Their paths are, however, different and opposing. The empirical sciences begin with individual and concrete phenomena and attempt to generalize their conclusions in an endeavor to expose the most basic and general laws that govern nature. The philosophy of nature works the other way around. It begins with the most general and fundamental principles and attempts to use them to constitute and explore the more specific aspects of nature.

Each of these two methods of inquiry has its limitations. Empirical science can only describe what there is, but not what there must be. Nor can empirical science justify the most general principles that underlie its inquiry, for example, the assumption that nature is indeed subject to laws. The philosophy of nature, on the other hand, cannot determine the nature of the phenomena that are revealed to us through experience. As mentioned earlier, even if conceptual analysis can determine the existence of, for example, a constant speed of determination in nature, which established the measurement of both time and space, the identification of any specific phenomenon in nature with that speed, for example, with the speed of light, is necessarily expressed by a contingent proposition, and hence extends beyond the boundaries of a priori research.

A complete understanding of nature hence requires both empirical and conceptual research. The goal of both sciences, which is a thorough understanding of nature, would only be achieved once the conclusions of the two methods of inquiry into nature come to mutual agreement and become integrated.

Unfortunately, while physics was established as a viable science centuries ago, the a priori research of nature has fallen behind and failed to contribute its part to what should be a joint endeavor. Although some philosophical ideas have played a part in the development of physics, the philosophy of nature as a research project has failed to progress like its empirical counterpart, and hence had little to offer in terms of insights and guidance. Hence, the most important immediate goal of the philosophy of nature should be its establishment as a science. It is my hope that this book contributes to the realization of this end.

Index

G. Yehezkel, *Conceptual Principles of Natural Philosophy*, Synthese Library 495, https://doi.org/10.1007/978-3-031-82108-0

M

N

O

S

T

The manufacturer's authorised representative in the EU is Springer
Nature Customer Service Centre GmbH, Europaplatz 3, 69115 Heidelberg,
Germany. If you have any concerns regarding our products, please
contact ProductSafety@springernature.com

Printed and bound by CPI Group (UK) Ltd, Croydon, CR0 4YY
30/09/2026
02225570-0001